Mathematics and Mathematicians

Volume 1

Mathematics and Mathematicians

Volume 1

P. Dedron and J. Itard

Translated from the French by J. V. Field

The Open University Press
In association with Richard Sadler Ltd

The Open University Press
12 Cofferidge Close, Stony Stratford,
Milton Keynes, MK11 1BY, England

First published 1959 by Editions Magnard, Paris under
the title *Mathématiques et Mathématiciens*.

First published in the English language 1974 by
Transworld Publishers Ltd., in association with Richard
Sadler Ltd.

Published in this edition by The Open University Press 1978
in association with Richard Sadler Ltd.

Made and printed in Great Britain

British Library Cataloguing in Publication Data

Dedron, Pierre
Mathematics and mathematicians.
1. – (Open University. Set books).
1. Mathematics – History
I. Title   II. Series   III. Itard, Jean
IV. Field, Judith
510′.9   QA21

ISBN 0 335 00246 3

Contents

Foreword

By Graham Flegg
Faculty of Mathematics, the Open University

At the beginning of 1973 I was appointed Course Team Chairman for the Open University's second level course, AM 289 *History of Mathematics*, due for first presentation in 1975. One of the first tasks to which the Course Team had to direct its attention was the selection of one or more suitable books which could be used as 'prescribed texts' around which the course itself could be designed. This was by no means an easy task. After studying a not insubstantial number of books, already available in English, I was led to the conclusion that none of these on its own met all the criteria upon which the Course Team's judgement needed to be based. These criteria can be broadly summarized as follows:

(a) The standard of mathematical and historical scholarship should be impeccable.

(b) The content should be intellectually accessible to readers with only a 'general school' level of mathematical background.

(c) The mathematical material should be presented in a general historical context.

(d) There should be a suitable balance between 'mathematics' and 'mathematicians'.

(e) The reader should be exposed to suitable primary source material.

(f) There should be some analysis in depth of suitable specific topics, such as 'written numbers', 'arithmetical calculations', 'solutions of equations', etc.

To find a book meeting all these criteria began to seem well-nigh impossible until the existence of *Mathematique et Mathematiciens* in the original French edition was drawn to my attention by Dr. D. T. Whiteside of Cambridge University. For this, I must record my sincere gratitude to him, and also to Judith Field for undertaking the lengthy task of translation and to Richard Sadler for agreeing to include the English version in the Transworld Student Library. The two-volume work now presented thus results from an original piece of very wise counsel and a substantial amount of very dedicated labour without which it is highly likely that this excellent work would never have been made available to the general English reader. The practising research mathematician of today may perhaps argue that by far the greater part of contemporary mathematics has been invented in the last hundred years and that to present a history of mathematics which excludes the more recent developments is to present only a small fraction of the history of the subject. However, it can also be argued that there is a real sense in which 'history repeats itself' and that to understand the way in which mathematical concepts develop it is not by any means essential to study the mathematics of recent years. Indeed, it may be argued further that a deeper and more balanced historical perspective can be obtained when the material under discussion does not include the contemporary or near-contemporary. This work presents the facts as far as they are known and within the constraints set by its original authors with clarity and authority. It does not attempt to interpret the facts in accordance with any particular view of history in general or of the history of mathematics in particular.* It does not progress far into the nineteenth century, and hence it satisfies the criterion that its content should be accessible to the non-mathematician. However, because the Authors wrote within this constraint, there is no detailed discussion of the invention or development of the infinitesimal calculus, and for this the reader interested in the technical matters associated with the calculus must look elsewhere.† Nevertheless, despite the absence of reference to the more recent developments in mathematics and to the technicalities of the mathematics of Newton and Leibniz, there is a more than adequate wealth of material in these two volumes to satisfy not only the enquiring general reader but also the serious student of the history of mathematics seeking a basis of sound scholarship upon which further and more specialized study can be built. Perhaps above all else, the Authors should be congratulated in introducing the reader to a substantial amount of primary source material which gives some fascinating insight into the minds of great mathematicians of the past, and without which no historical study of mathematics can be seriously undertaken.

MILTON KEYNES
1974

---

* For a stimulating interpretation of the facts relating to the history of number and of geometry the reader is referred to *The Evolution of Mathematical Concepts*—R. L. Wilder—Transworld 1974 (also a prescribed text for the Open University History of Mathematics Course).

† Two particularly useful books are: *The Origins of the Infinitesimal Calculus*—M. E. Baron—Pergamon Press, and *The History of the Calculus and its Conceptual Development*—C. B. Boyer—Dover.

Authors' Preface

The present work does not claim to be even a brief history of mathematics or mathematicians. It offers varied and fairly straightforward material in the hope of provoking a certain amount of reflection about the birth and development of the mathematical sciences.

Mathematics arose from the need for a system for counting and for calculating areas and volumes, but it has over the centuries become less concerned with practical matters and has turned instead towards logic and pure intellectual speculation.

In the first part of our book, we have tried to show some of the intentions, the hesitations and the doubts in the minds of the greatest thinkers of ancient times and of the Middle Ages which led to that flowering of new ideas and more sophisticated methods which came about with the Renaissance and in the works of the great mathematicians of the seventeenth and eighteenth centuries.

The second part of our book deals with the history of methods, and with several particular problems, some of which were definitively solved long ago while others, many of them famous, intrigued and obsessed the most learned scholars for more than three thousand years until they were finally proved to be insoluble by the geometrical means prescribed.

Trisecting the angle and doubling the cube were the first and the simplest of the problems which turned out not to be soluble with only a straight edge and compasses. However, these two were not 'transcendental' problems like the famous squaring of the circle, which has passed into common parlance to designate an impossibly difficult task.

To have described modern mathematics, even in the simplest manner, would have been very cumbersome, and since we wished to confine our attention to elementary mathematics, we decided, with regret, that we must end our survey at the beginning of the nineteenth century. We should have been unable to describe in adequate detail the immense work of the many eminent scholars, such as Chasles, Evariste Galois, Henri Poincaré, Abel, Steiner, Hilbert, and many others.

Towards the middle of the nineteenth century, mathematicians began to take an interest in almost entirely new subjects: group theory, topology, axiomatic systems and non-Euclidean geometries. Discoveries were rapid, striking and very significant.

Those of our readers who are accustomed to mathematical language, to the strictly defined meaning attached to words (whether invented for their purpose or borrowed from ordinary usage) will notice that even in relatively recent writings special notations and symbols are rare. Expressions are clumsy, the terms are imprecise, and there are numerous periphrases which would be severely criticised in the work of a student of today.

The curious might be interested to translate into modern technical terms some of the proofs given by Euclid or even by Fermat. It might be an absorbing exercise to search for the correct term and the most concise and cogent argument.

We might even suggest that the bolder spirits turn their attention to texts in other languages: translating a passage of Latin, for example, might be profitable in more than one respect, even though its style might have little of the literary quality of Cicero or the historical acuity of Tacitus.

We also hope that, having accompanied us through the stages of this brief history of the mathematical sciences, some of our readers will find it has caused them to reflect not only on historical and linguistic matters but on philosophical concepts also.

It is perhaps not without interest to note that, for example, geometry was highly developed in Antiquity, and Euclid's *Elements* have for centuries been the basis for all our teaching in the subject, whereas in other branches of mathematics even the most subtle and acute of scholars ran up against difficulties which were long considered insurmountable and often described as paradoxes. Notable examples are irrational numbers, infinitesimals, convergent series, limits and continuity of functions. None of these was defined or properly studied before the eighteenth or nineteenth centuries.

We do not wish to follow the wrong-headed example of those who believe they can contrast a 'spirit of geometry' with a 'spirit of analysis' but we are compelled to take note that the development of geometry and that of analysis were very different; the gap of so many centuries which lay between their arriving at maturity must be compared with what we see in disciplines other than mathematics.

The twentieth century probably holds further surprises. A few years have seen a complete revolution in physics and chemistry, brought about by the study of atoms. Many young mathematicians are now campaigning for a complete renewal of the logical structure of mathematics: they are concentrating on topology and on axiomatic systems, their nature, number and interdependence. There is an increasing tendency towards complete abstraction and towards theories whose scope embraces 'sets' of very varied and even apparently disparate geometrical and algebraic elements.

Yet again different methods and new ideas have required mathematicians to introduce a very large number of new technical terms which can easily baffle the layman: *time and patience will make them familiar*.

Mathematics of the most rigorous and apparently the most abstract kind has for a long time been essential to the development of all the sciences: physics, chemistry, biology, statistics, economics, and even psychology and ethics.

Our aim has been to interest our readers in the history of the subject and we hope to inspire some of them both to study it and to admire its process of continual creation.

PIERRE DEDRON

Acknowledgements

We gratefully acknowledge our indebtedness to the following publishers for their permission to quote from translations of primary sources material: Cambridge University Press, from *The Thirteen Books of Euclid's Elements*, translated by Sir T. L. Heath (all quotations from the *Elements* are from this source unless otherwise stated); and from *The Mathematical Papers of Sir Isaac Newton*, translated and edited by D. T. Whiteside: The Clarendon Press, Oxford, from *A History of Greek Mathematics* by Sir T. L. Heath (all quotations from Archimedes, Apollonius and Euclid (other than the *Elements*) are by kind permission of The Clarendon Press): Encyclopedia Britannica, from Ptolemy's *Almagest*, translated by R. Catesby Taliaferro: William Heinemann Ltd., from *The Mathematical Collection of Pappus*, translated by Ivor Thomas.

We are also grateful to the Bibliothèque Nationale, Paris, The Mansell Collection, London, and Photographie Giraudon, Paris, for the illustrative material used in both volumes of this book.

Publisher's Note

In general we have adopted a policy of putting quotations from the works of mathematicians into a type smaller than the 'text' type, so as to make them readily distinguishable. Commentaries by historians have been put in quotes, as have brief quotations from mathematicians where to extract them would have risked breaking the flow of the narrative.

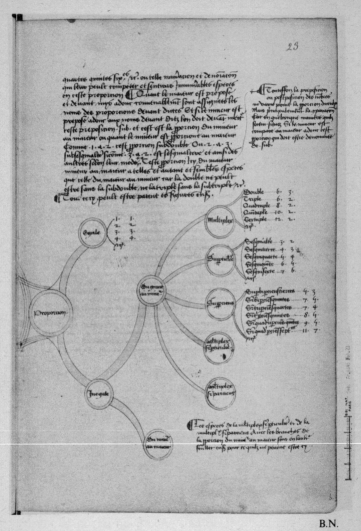

PLATE I. A fifteenth century diagram representing the nomenclature of numerical ratios. From *Triparty en la Science des Nombres* (1484) by Nicolas Chuquet.

1. Bird's Eye View

Multi pertransibunt et augebitur scientia.

As long as mankind has existed there have always been artists, painters and carvers. Their works, striking in their truth and elegance, are still to be found on the walls of the caves along the river valleys of France and Spain.

There must also have been storytellers, poets and musicians whose traces are harder to follow, though they can perhaps be found in the age-old traditions of folklore, which with the invention of writing were incorporated into the first literature.

Are we then to believe that among these remote ancestors of ours there were no spirits who enjoyed abstract thought, or speculated about numbers and shapes, or appreciated the beauty of the pattern of the stars in the night sky, of the flowers formed by the frost, of the sharp edges of crystalline rocks?

Many peoples have legends of these learned men of long ago. Josephus, historian of the Jewish War, says of them:

It is to their intelligence and their labours that we owe the science of astrology; and, since they had understood from Adam that the world would be destroyed by water and by fire, their fear that this knowledge might be lost before men could be told of it led them to build two columns, one of brick and the other of stone, on which they carved the knowledge they had gained, so that if a flood were to destroy the brick column the stone one would remain to conserve for posterity the memory of what they had written on it. Their foresight triumphed; and I am told that the stone column can still be seen today in Syria ... Also, our forefathers were particularly dear to God, and since they were the work of His hands and since the food they ate was such as to preserve life, God lengthened their lives on account of their virtue and also so that they might perfect the sciences of geometry and astronomy which they had discovered: this they could not have done if they had lived less than six hundred years, because only after six centuries is the Great Year complete.

Legends always contain some grain of truth and the latest research into the history of mathematics has shown that the origins of science must be sought in very remote antiquity.

There are surviving examples of numerical notation which date back to the beginnings of writing in the third millenium BC. Documents which can be dated to about 2000 BC, that is, to a time as long before the birth of Christ as our own is after it, show that both the Egyptians and the Babylonians had sophisticated systems of mathematics.

But such documents are rare. They give us precise information about particular points, so they are good snapshots recording certain historical details, but they are not panoramic photographs, nor, still less, are they historical films which allow us to reconstruct the life and intellectual climate of a period.

For the moment, therefore, we must be content with incomplete and tentative information when we refer to these ancient times. So far as the Egyptians are concerned, we know about their numerals, their method of calculation, and the way they dealt with some first degree algebraic problems and some calculations of areas and volumes. The exact interpretation of some of these calculations is still in doubt. We know rather more about Babylonian mathematics, but our knowledge is too fragmentary to allow us to compare the scientific levels of the two civilizations. In the present work we shall describe the sophisticated system of numerals used by the Babylonians, Babylonian algebra (both first and second degree) and shall discuss the way the Babylonians used the Theorem of Pythagoras.

We still know nothing at all about the levels of mathematical development of those of the other civilized peoples of the Middle East whose writings can be deciphered, or who, like the Jews, are known to us through a more or less unbroken tradition.

In this, as in many other spheres, the Greeks were the initiators in the Western world. Little is known of their early work and in the chapter entitled 'The Mists of Legend' we merely mention some aspects of it, without any historical criticism. However, despite the scarcity of documents, or even the complete lack of them, we do know that the Greeks of the fifth and fourth centuries BC had an extensive knowledge of mathematics and, above all, an intellectual grasp of the subject. Since our information comes from philosophers who were more or less mystical and poetic, we tend to see Hellenic mathematics in a rather sophistic light. Nevertheless, we should remember that every surviving document is precious not only because of the facts it purports to relate but also because of what it tells us about the writer of the document.*

From this time onwards, particularly in the fourth century BC, Greek mathematical work goes beyond what we now consider as the domain of elementary mathematics. The problems of doubling the cube and of trisecting the angle, and certain interpolation problems, led to cubic and

---

* Letter from Gauss to Schumacher, 1 November 1844; 'You can see the same sort of thing [mathematical incompetence] among to-day's philosophers—Schelling, Hegel, Nees von Esenbeck, and their followers. Don't their definitions make your hair stand on end? Read the explanations given by the great men of the past, Plato and others (I except Aristotle). And the same with Kant, he's often not much better: to my mind, his distinction between analytic and synthetic propositions is one of those things that are either banal or wrong.'

quartic equations. The problem of squaring the circle is, however, the first example of a transcendental problem, and the discussion of the angle of contingency which first arose a little before 300 BC, opened up a line of research which was not brought to a conclusion until the end of the nineteenth century AD.

It was probably at the end of the fourth century that the *conic sections* were discovered. More complicated curves, algebraic but crude, are to be found in the work of Archytas of Tarentum, and plane transcendental curves, such as the quadratrix, appear in the work of Dinostratus.

The third century BC, at the start of the Hellenistic period, was the great age of ancient mathematics.

We shall discuss this period at some length but, although we can give an almost exhaustive account of Euclid's *Elements*, we shall have to omit his fifth and tenth books, which, particularly the fifth, deal with mathematical problems beyond the scope of the present work.

Some of Archimedes's mathematics has been incorporated into the school syllabus, but only at the sacrifice of some of its rigour and hence of some of its educational value.

Apollonius, for his part, deals with conic sections in a manner much closer to that of specialist mathematics than to that of present-day elementary teaching.

Euclid, Archimedes and Apollonius dominated mathematics until the beginning of the eighteenth century AD but Ancient mathematics in fact continued after their time. Ptolemy, Diophantus and Pappus also contributed much that was to be of value to Western scholars of the sixteenth and seventeenth centuries.

The Graeco-Roman period saw the beginning of the long era when knowledge was merely transmitted. We describe the broad outlines of this period in Chapter 5.

A slow revival, which has not received sufficient attention, began in the thirteenth century. We somewhat arbitrarily regard this period of revival as coming to an end at the close of the fifteenth century. Modern mathematics, directly descended from that of the Greeks, begins with the sixteenth century.

At the end of the fifteenth century Nicolas Chuquet had invented a notation for exponentials, which is used later on in the work of Bombelli, Stevin, Albert Girard and Descartes. Negative numbers came into use, although they were only accepted very slowly. But the great age of elementary algebra came in the sixteenth century: on the one hand there was the Italian school, centred on the University of Bologna, whose work led to solutions for cubic and quartic equations and thus to the discovery of complex numbers, while on the other hand great advances in the use of symbolic notation were made in Germany, Flanders and Italy, and particularly by Vieta in France.

At this opportune moment the manuscripts of Diophantus were discovered and translated into Latin. Vieta, well read in the Greek classics, was particularly active in introducing the *literal calculus*, the use of letters to represent unknown quantities. This gave mathematics a universal language and proved to be a magnificent intellectual tool.

The early seventeenth century, a period German historians refer to as the age of the Baroque, saw great progress in mathematics. By this time, most of Ancient Greek mathematical work was known and had been translated and absorbed, particularly the work of Euclid, including Book X. However, Book V was not to be understood for a long time yet.

The more elementary part of Archimedes's work had also been understood and it inspired Luca Valerio and Stevin (among others) to look for simple and elegant ways of determining centres of gravity.

Apollonius's four first books on conics were known, and Maurolico and Kepler used them in their work on the refraction and reflection of light.

Pappus's *Mathematical Collection* had just been translated and it encouraged an analytical approach to geometry. The outstanding contributors in this field, apart from the pioneers Vieta and Marino Ghetaldi, were Fermat and Descartes.

Trigonometry, which had long been the undisputed province of Ptolemy, began to take on its modern form, particularly with the work of Vieta. In calculating tables it became more usual to divide the radius into hundredths, and sixtieths were less used. Stevin published his *Dixme*, which recommends the use of decimal fractions. The problem of squaring the circle, which had been taken up again in the fifteenth century, now became the subject of increasingly elaborate decimal calculations, and stimulated work on infinite algorithms.

Napier and Bürgi invented *logarithms*. Exponential functions thus took their place alongside trigonometrical ones.

Kepler's work in optics and astronomy and Galileo's work in mechanics showed a completely new use of mathematics.

Fermat, Descartes and Wallis laid the foundations of analytical geometry.

Kepler, Cavalieri, Roberval, Fermat, Gregory of Saint-Vincent, Toricelli and Pascal rediscovered Archimedes's analytical methods for themselves without having an explicit knowledge of his fundamental work on the subject, and in their very different styles all contributed to the invention of a primitive form of the integral calculus known as the *geometry of indivisibles*.

Lord Brouncker and Wallis devised new algorithms which made use of infinite products and continued fractions.

Desargues, followed by Pascal and later by La Hire, laid the foundations of projective geometry.

Fermat, Descartes, Roberval and Torricelli invented ways of finding tangents to curves. Fermat and Descartes employed the methods of analytical geometry, while Roberval and Torricelli used kinematic methods which look forward to vector analysis. Barrow made explicit the connection between these new problems and problems involving indivisibles, a connection already to be seen in the work of Fermat, Roberval and Torricelli. Pure algebra had already attained a high level, thanks to the work of Bombelli and Vieta, and further remarkable progress is to be seen in the works of Albert Girard, Harriot, Oughtred, Fermat and Descartes. The analysis of rational numbers, based on the work of Diophantus, although now an intellectual backwater, was then an important part of the mainstream of mathematical thought and stimulated a great deal of work. Some knowledge of it is essential to an understanding of seventeenth century analysis.

Fermat developed his own work from a thorough and critical examination of that of Diophantus, and thus, single-handed, laid the foundations of number theory. Together with Pascal he established the bases of the *calculus of probabilities*, and then, by his analysis of Descartes's work in optics, the bases of the *calculus of variations*.

This period, which, like any transitional period, is difficult to delimit exactly, was extremely important for the problems it raised and solved. It drew its inspiration from Ancient Greece, but also from the sixteenth century and often, too, from the scholasticism that lingered on at least until the time of Descartes. The mathematical vocabulary of the time was muddled, verbose and anarchic: there were terms derived from the Latin of the Scholastics and from the Latin of the Humanists, and many terms were invented from the Greek by Vieta, or borrowed by engineers from the language used by craftsmen. Notation, often inadequate, underwent continual changes. The mathematicians themselves came from various backgrounds: there were university teachers, particularly in Italy, Germany, Flanders and England; there were engineers (Bombelli, Stevin and Albert Girard), or architects (Desargues); in France there were lawyers (Vieta and Fermat); or sons of well-to-do families (Descartes and Pascal).

Men such as Wallis, Barrow, James Gregory and particularly Huygens mark the transition to a new age which in England began with Newton and on the Continent with Leibniz. Mathematicians still for the most part wrote in Latin, though French began to predominate after the time of Descartes, and the work of these new men has considerable uniformity in its mathematical style and notation.

Vieta's principles were at last accepted. Problems were no longer categorized as being either geometrical or numerical, but were all considered to be within the scope of generalized analysis. Terminology became more precise: the words 'series' and 'convergence' were introduced in England, and were used in senses increasingly like their modern ones, while Leibniz and Jean Bernoulli used the word 'function' instead of the imprecise expressions of various origins which had previously been used to describe an ill-defined idea.

Calculus advanced particularly rapidly and for a time it seemed to include all of mathematics. The term covered various infinitesimal techniques: *differential calculus* (Newton's fluxions), *integral calculus* (for Newton a return from fluxions to fluents), the study of *differential equations* and the *calculus of variations*.

The work of the brilliant period that was to follow unfortunately lies beyond the scope of this book.

Two of Newton's immediate disciples, Taylor and Maclaurin, made particularly distinguished contributions during this time. Leibniz's disciples included Jacques and Jean Bernoulli, and in France there was the rather less important figure of Guillaume de l'Hôpital. The most significant work was that of the Bernoullis: their applications and extensions of Leibniz's techniques proved to be exceedingly fruitful. The genius of their brilliant pupil, Leonard Euler, was to dominate every branch of eighteenth century mathematics. In France, however, there was a lull at the end of the seventeenth century, the only significant work being that of L'Hôpital, Varignon and Rolle, but the eighteenth century saw the appearance of Clairaut, D'Alembert and then Laplace, whose work was to combine Leibniz's analytical techniques with Newton's monumental work in mechanics and astronomy. Lagrange, who was born in Italy but was of French descent on his father's side, lived for a long time in Berlin, but finally established himself in Paris, becoming a Frenchman by adoption. Almost from the first, his mathematical work was written in French, and his textbooks, written towards the end of his life, were, with those of Monge, the most important formative influences on the great French mathematicians of the nineteenth century.

Monge was to discover descriptive geometry, and to exert great influence on the French geometers of the following century. He also founded the École Polytechnique. As a mathematician Lacroix showed less originality than Monge but his work described the new developments of the eighteenth century in a systematized and simplified form, while Legendre, whose solid scholarship is not always adequately recognized, followed up Fermat's and Euler's work on number theory, and carried out original work which laid the foundations for the study of elliptic functions. His diligent though not completely successful work was eventually to lead to non-Euclidean geometries. Work was also done in this field by Lambert, a Frenchman who was born in Mulhouse in Switzerland and spent the greater part of his academic life in Berlin. Most of his work was written in German.

The outstanding pure algebraists of the eighteenth century were Waring in England, and Vandermonde and Lagrange, among others, on the Continent.

The mathematicians of this time were, however, more interested in proving new theorems than in finding rigorous proofs of old ones.

In this respect, the scientific spirit had declined since the time of Newton, and several paradoxes appeared as a result. A salutary reaction is already noticeable in the work of Lagrange, and it developed to the full in the work of Gauss in Germany and of Cauchy in France. Both these scholars were prolific, and their rigorous work leads us on into the nineteenth century, when a fine school of geometry came into being in France, exemplified by men such as Poncelet and Chasles, and later, in Germany with, for example, Möbius, Plücker and the Swiss, Steiner, the perfect type of the pure geometer.

During the sixteenth century the Italians had made remarkable advances in algebra, but since then no further progress had been made in the solution of equations. Euler and, more particularly, Lagrange eventually identified the difficulties. Vandermonde made an intensive study of certain types of equation and the young Gauss's work in the same field showed that the equation $x^{17} - 1 = 0$ reduced to a finite set of quadratic equations: so the regular polygon with seventeen sides could be inscribed in a circle by the use of only a straight edge and compasses. This was the first important advance in the study of regular polygons to be made since the fourth century BC. However, Ruffini, Abel and Galois proved that the method of solving equations that had been used up to then was only applicable as far as quartic equations and could not be used for the general quintic. Further, they showed, that no finite series of additions, subtractions, multiplications, divisions and extractions of roots of any degree could lead to a solution of the general quintic equation.

This important result, from which Wantzel immediately concluded that the problems of doubling the cube and trisecting an angle could not be solved with a straight edge and compasses, was the first of the series of negative theorems which were to be proved in the nineteenth century.

We have already mentioned that at this time Legendre and Lambert worked on problems connected with Euclid's parallel postulate.* Gauss had obtained some interesting results from his work on this postulate, but had not published them. It was left to Bolyai and Lobachevsky to establish beyond doubt that Euclid's parallel postulate could not be deduced from the other axioms of Greek geometry. Investigations of non-Euclidean geometries were important in the development of the study of the foundations of mathematics, and with them there appeared a new branch of mathematics: the study of axiomatic systems. The most important contribution to this was that made by Hilbert towards the end of the century.

At the end of Chapter 7, Vol. 2, we shall mention Lindemann's proof of the impossibility of squaring the circle. We have been compelled to omit the history of the dispute about the angle of contingency, which first appears in Book III of Euclid's *Elements*, because the considerations involved are too advanced. This dispute ended at the close of the nineteenth century with the work of Cantor, from which Emile Borel and Henri Lebesgue derived new and extremely important techniques whose contribution to

---

* See p. 70.

twentieth-century mathematical thought was comparable only to that made by the discussion of Euclid's postulate.

Lagrange and Cauchy's work on algebra marks the first appearance of an idea which was to come to fruition in the work of Evariste Galois: the idea of a group of transformations. A Danish mathematician, Sophus Lie, extended this work to geometry and analysis by studying continuous groups. The philosophical writings of Poincaré show the far-reaching consequences of this idea in all branches of mathematics. In the twentieth century Elie Cartan was to make it his speciality.

The idea of a determinant arose from the work of Leibniz in the seventeenth century, that of Cramer, Bezout and Vandermonde in the eighteenth, and of Cauchy in the nineteenth. The work done by Cardan and, more particularly, by Bombelli in the sixteenth century led to an important generalization of the concept of number: the discovery of complex numbers. Such numbers were used during the seventeenth and eighteenth centuries, when their properties were not yet properly understood. They were thoroughly explained by about 1800, and many writers, the most important of whom was Argand, developed geometrical representations of them. In the hands of Gauss and Cauchy complex numbers became indispensable tools in the study of analytic functions.

The word 'function' was introduced into the vocabulary of mathematics by Leibniz and Jean Bernoulli about 1700. At that time the known functions were, on the one hand, the functions we now call algebraic (for which the function values y and the variable x are related by an equation of the form $P(x, y) = 0$, P being a polynomial in x and y), and, on the other hand, trigonometric functions, variously derived from Greek mathematics. There were also exponential functions, which, in the form of logarithms, had been known since about 1600. Many more functions were introduced in the eighteenth century, among them the elliptic functions. Legendre was the first to study these systematically, and considerable advances were later made by Abel and Jacobi. Hermite's work was also important.

The crowning achievement of the nineteenth-century work in this field was Henri Poincaré's brilliant extension of the theories of elliptic functions to cover new entities, which he called *Fuchsian functions*, after the mathematician Lazarus Fuchs.

The geometrical representation of complex numbers led some mathematicians to make further generalizations of the idea of number. Hamilton's quaternions were an instance in point. Among the results which have now become part of elementary mathematics we should mention *vector analysis*, which has proved useful because it is very simple and can be grasped intuitively. The more sophisticated types of analysis associated with matrices and tensors, which are very important in their applications to physics, are of similar origin.

One of the aims of this present work is to show that mathematical knowledge has only reached its present high level through the labours of many generations. In the words of Francis Bacon and Pierre Fermat:

Many shall pass, and knowledge shall gain by them.

2. The Mists of Legend

*Che non men che saper dubiar m'aggrada**
Dante

We know very little about Egyptian mathematics, and almost nothing at all about the mathematicians who lived in the Nile valley in the third millenium BC. Only three or four papyri have survived, of which two, the Rhind Papyrus and the Moscow Papyrus, have been studied in detail.

Who were these mathematicians? Some fifteen hundred years later Democritus, a Greek, called them 'harpedon-aptes', stretchers of cords, and Aristotle spoke of priests who occupied their leisure with mathematics. Geometry could have arisen in response to the needs of surveyors and arithmetic in response to those of merchants. The Greeks, however, attributed the invention of arithmetic to the Phoenicians, who were pirates and traders.

In fact, from the documents that have come down to us, which are merely school manuals, it seems clear that the ancient Egyptians had a thorough grasp of the rules of arithmetic and of the rules for measuring areas and volumes. For instance, they knew how to find the area of a circle and the volume of a truncated pyramid.

It is possible that the leisured classes, well-grounded in the everyday techniques of practical geometry and

---

* To be in doubt pleases me as much as to know.

computation, might have used their knowledge as a basis for mathematical investigations, but the only evidence we have is a brief and vague reference by Aristotle.

We are better informed about Babylonian mathematics. Mesopotamia and the surrounding regions were battle-grounds for thousands of years, and it is to this constant warfare that we owe most of our information.

Scribes wrote on soft clay tablets, roughly the size of a man's hand, using a metal stylus. Mesopotamian children, like their contemporaries among the literate classes of Egypt, attended schools, where they learned reading, writing, arithmetic and vocational skills with the aid of these tablets. Kings such as Assurbanipal assembled large libraries of them. The tablets, originally fragile, were baked hard like bricks when, in time of war, cities were sacked and set on fire, and the dry climate of Asia Minor helped to preserve them. Documents, some worthless and some valuable, thus lay buried for thousands of years in the great mounds formed by the ruins of abandoned cities such as Niniveh and Susa.

It is now about a hundred years since such tablets (Plate II) were first unearthed, and they have mainly been deciphered within the last fifty years. As more and more tablets were deciphered it gradually became clear that the Babylonians knew a great deal more about mathematics than had previously been supposed. They used subtle techniques of computation, they had a sophisticated system of algebra, and their geometrical work, though not equal to that done in Hellenistic Greece, was far from negligible. We shall return to the subject of Babylonian mathematics later, but of the mathematicians themselves we know nothing.

The Greeks recorded many stories about their earliest scientists, but none stands up to historical criticism.

We shall begin with Thales.

B.N.

PLATE II. Collection of problems on a Babylonian tablet in the British Museum. (From O. Neugebauer, *The Exact Sciences in Antiquity*, Copenhagen, 1951.)

He was one of the Seven Sages.* Many stories are traditionally told of him but all, or almost all, are legendary.

He is said to have lived seventy, eighty, ninety or even a hundred years (the last is according to Lucian) and to have been born about 640 BC in Phoenicia, according to some, or in Miletus, according to others. We know for certain that Plato and Aristotle quote him. He is said to have been a merchant, and the story-tellers have contributed the following legend:

One day when Thales was leading a caravan, a mule loaded with salt fell into the water while crossing a ford. After it had scrambled out it noticed its load had become lighter, so at the following ford it deliberately rolled in the water. To cure it of the habit, Thales loaded it with sponges. When the mule tried its trick again its load became not lighter but much heavier. This is one of the fables told by Aesop, who was possibly a contemporary of Thales.

Aristotle says that Thales was a clever speculator who bought up all the oil-presses in the district, and another tradition describes him as a skilful engineer who turned a river from its course so that a ford could be made.

He is said to have travelled to Egypt where the priests initiated him in their ancient wisdom. Paul Tannery considers the legend to be rather exaggerated since a foreign merchant would surely have found it difficult to gain the confidence of such an exclusive caste. What Thales could have learned in Egypt would have been no more than the elementary ideas current among the lay public.

There is a story that he astonished the Pharoah by measuring the height of the pyramids by means of their

---

* Plato, *Protagoras* 343a: Thales of Miletus, Pittacos of Mytilene, Bias of Priene, Solon, Cleobulus of Lindos, Myson of Chene, Chilon of Sparta. Thales is the only one of the seven who is a philosopher and a scientist. Moreover, the list of the Seven Sages varies from one author to another.

shadows. This is an elementary geometrical problem involving the use of similar triangles. The Rhind papyrus contains several other problems to do with the pyramids, involving the idea of a *seqt*, which is the inverse of the gradient of the faces, that is the cotangent of the angle they make with the horizontal.

An Old-Babylonian tablet, No. 85 120 in the British Museum [1], contains several problems of this sort: a staircase of equal steps is of height H and horizontal length L. Each step is of height h and width l. The problem is to find one of the four measurements given the other three. The solution clearly involves using proportionality.

$$H:L = h:l.$$

The Pharoah's astonishment at Thales's calculations must therefore be considered entirely a matter of legend.

We need hardly add that our 'Thales's Theorem' is not in fact due to Thales. Moreover, this theorem, which states that the segments cut off from transversals by parallel lines are proportional to one another, is far from being as ancient and venerable as Pythagoras's Theorem. The Egyptians and the Babylonians accepted it, but not as a proposition that required proof: it was just an obvious fact, like all ideas of similarity. Pythagoras's Theorem, on the other hand, is not obvious, and had to be discovered.*

We should keep the name 'Thales's Theorem', but only as an act of homage to one of the first learned mathematicians, and bearing in mind that this title is very recent in origin.

In Definition 17 of Book 1 of the *Elements* Euclid states:

The diameter of a circle is a straight line through the centre, cut off at either end by the circumference of the circle; it thus divides the circle into two equal parts.

---

* See Chapter 4, Vol. 2.

Proclus adds in his commentary:

The famous Thales is said to have been the first to prove that the circle is bisected by the diameter [2].

It is hard to believe Proclus's statement, since three centuries after Thales we find that Euclid, one of the most rigorous geometers of antiquity, does not prove this property but merely incorporates it in a definition.
Proclus also writes of the isosceles triangle:

We are indebted to old Thales for the discovery of this and many other theorems. For he, it is said, was the first to notice and assert that in every isosceles triangle the angles at the base are equal, though in somewhat archaic fashion he called the equal angles similar [3].

This seems more plausible. Nowhere in the early Babylonian or Egyptian literature do we find a reference to angles. We can thus, temporarily, at least, acknowledge that the Greeks may have been the first people to concern themselves with the study of angles, and credit Thales with some part in it.

We may, perhaps, form an idea of how Thales might have approached the problem of the angles at the base of an isosceles triangle by looking at how Aristotle treated the subject in the fourth century BC [4].

Aristotle considered a circle with its centre at the vertex of the isosceles triangle and with radius equal to the two equal sides of the triangle, A and B (Fig. 2.1).

The angle AC is equal to the angle BD since they are

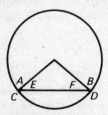

Fig. 2.1

'angles of the semicircle' i.e. right angles. The angle C is equal to the angle D since they are angles of the same segment. If from two equal quantities we subtract two equal quantities the remainders are equal, so angle E = angle F.*

At the beginning of Book III of the *Elements* Euclid considers angles bounded by a straight line and an arc of a circle. Definition 7 states:

An *angle of a segment* is that contained by a straight line and a circumference of a circle.*

But Euclid's own proof that the angles at the base of an isosceles triangle are equal depends only on equalities and involves only angles bounded by straight lines. It is completely rigorous.

Proclus also credits Thales with having discovered that opposite angles at a vertex are equal [5] and says that Eudemus, a pupil of Aristotle and the first historian of mathematics, reported that Thales discovered the following theorem [6]: Two triangles which have one equal side and two equal angles are congruent. Eudemus is said to have added that this theorem is needed to calculate the distance of boats at sea by a method which Thales is said to have invented.

We shall not concern ourselves here with Thales's contribution to philosophy, ethics or astronomy, but it should be mentioned that he owed much of his fame to having predicted an eclipse. According to Montucla: 'This eclipse was the one which occurred at the moment when Cyaxares, King of the Medes, and Aliathes, King of the Lydians, were about to join battle. According to Riccioli's calculations it took place in 585 BC, and this is confirmed by Pliny, who dates it to the fourth year of the forty-eighth Olympiad' [7].

---

* Aristotle made use of *mixed* angles, that is, angles contained in a straight line and a curvilinear arc. Thus angles AC and BD are angles between radii and the circumference, and angles C and D between the chord CD and the circumference. H.G.F.

In fact, although opinion is generally in favour of 28 May 585 BC (by the Julian calendar), two other dates are also possible: 21 July 597 BC and 30 September 610 BC. Herodotus gives an account of the battle. The eclipse of 597 according to modern calculations took place in the morning and does not fit in with Herodotus's account. According to the same calculations, which are not, and cannot be, exact, the eclipse of 610 was not total in Cappadocia, where the battle took place, though it was total in Armenia.

The fact that the eclipse took place, that it took place at the beginning of the battle, and its effect on the course of the battle are all well-established. But could Thales have predicted it?

It was only considerably after the sixth century, in fact after the time of Alexander, that Chaldean astronomers could give reasonably accurate predictions of eclipses of the moon. They were never able to predict that an eclipse of the sun would be seen in a particular region.

Thales's prediction has been explained by his having supposedly made use of the Babylonian period known as the Saros. Unfortunately, the Saros is an astronomical period which was never in fact used by those who are credited with having discovered it. The idea of the Saros first originated in the mind of Edmund Halley, as an interpretation of a passage in Pliny, and it was published in the Philosophical Transactions of the Royal Society in 1691. It was severely criticized by Le Gentil in 1766, but, unfortunately, was accepted by Montucla in 1758, and later became an article of faith [8].

Reduced to its proper proportions Thales's prediction, as described by Herodotus, must be seen as no more than the prediction that daylight would fail on one day in the year of the battle.

Callimachus, an Alexandrian poet of the third century BC, recounts that when the son of Bathycles brought Thales the cup which was to be given to the worthiest of the Seven Sages, he found the old man engaged in drawing on the ground a figure that had been discovered by Euphorbus, the Phrygian, the first man to draw scalene triangles in a circle.

We know nothing at all about Euphorbus beyond Callimachus's assertion that he studied geometry before the time of Thales.

There is a tradition that Pythagoras, who believed in the transmigration of souls, claimed that in a previous life he had been called Euphorbus, but we cannot be sure whether he meant the legendary mathematician or the young hero who died in the Trojan ranks after having wounded Patroclus.

Pythagoras, a generation younger than Thales, seems to have lived in the second half of the sixth century BC. He was born in Samos and it is sometimes claimed that he was taught by Thales and by his pupil Anaximander. After much travelling Pythagoras finally settled in Magna Graecia (Southern Italy), where he became involved in the movements of political and social unrest. He founded a sect whose outlook was élitist, mystical and religious. Its adherents lived as a community and held their goods at least partly in common. They followed rigid rules and abstained from eating certain foods, such as beans and meat. Abstention from meat is connected with belief in the transmigration of souls, and it is interesting to note that Buddhism, which has several characteristics in common with Pythagoreanism, in fact arose in India at the time of Pythagoras.

The Pythagorean communities met with opposition, some of it bloody, and the destruction of the community of Croton, where Pythagoras himself is said to have met his death, put an end to the sect's political activities.

The Pythagoreans' most important achievements were in the domain of science, but since his followers always ascribed their discoveries to their master, Pythagoras, it is now almost impossible to distinguish between his work and theirs.

Euclid's discussion of odd and even numbers in Book IX of the *Elements* is possibly derived from some of the early work of the Pythagorean school.

Among the Definitions at the beginning of Book VII of the *Elements* we read:

1. An **unit** is that by virtue of which each of the things that exist is called one.*
2. A **number** is a multitude composed of units.
6. An **even number** (ἄρτιος ἀριθμός) is that which is divisible into two equal parts.
7. An **odd number** (περισσὸς ἀριθμός) is that which is not divisible into two equal parts, or that which differs by an unit from an even number.

The very elementary propositions to be found in Book IX may perhaps date back to Pythagoras and his immediate disciples. They are proved in a very archaic manner which is not logically rigorous:

PROPOSITION 21. *If as many even numbers as we please be added together, the whole is even.*

For let as many even numbers as we please, *AB*, *BC*, *CD*, *DE*, be added together;
I say that the whole *AE* is even [Fig. 2.2].

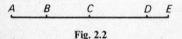

Fig. 2.2

For, since each of the numbers *AB*, *BC*, *CD*, *DE* is even, it has a half part;
[VII. Def. 6]
so that the whole *AE* also has a half part.
But an even number is that which is divisible into two equal parts; [*id.*]
therefore *AE* is even.

Q.E.D.

---

* Iamblichus claims the Pythagoreans used a different definition: unity (Μονὰς) is the boundary between number and parts of a number (μόριον).

PROPOSITION 22. *If as many odd numbers as we please be added together, and their multitude be even, the whole will be even.*

For let as many odd numbers as we please, *AB*, *BC*, *CD*, *DE*, even in multitude, be added together;
I say that the whole *AE* is even. [Fig. 2.3]

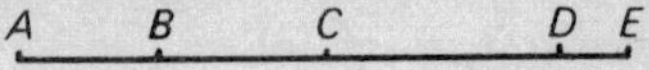

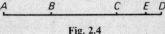

Fig. 2.3

For, since each of the numbers *AB*, *BC*, *CD*, *DE* is odd, if an unit be subtracted from each, each of the remainders will be even;   [VII, Def. 7]
so that the sum of them will be even.   [IX. 21]

But the multitude of the units is also even.
Therefore the whole *AE* is also even.   [IX. 21]
                                                                        Q.E.D.

PROPOSITION 23. *If as many odd numbers as we please be added together, and their multitude be odd, the whole will also be odd.*

For let as many odd numbers as we please, *AB*, *BC*, *CD*, the multitude of which is odd, be added together;
I say that the whole *AD* is also odd [Fig. 2.4].

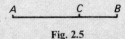

Fig. 2.4

Let the unit *DE* be subtracted from *CD*;
therefore the remainder *CE* is even.   [VII. Def. 7]
But *CA* is also even;   [IX. 22]
therefore the whole *AE* is also even.   [IX. 21]
And *DE* is an unit.
Therefore *AD* is odd.   [VII. Def. 7]
                                                                        Q.E.D.

PROPOSITION 24. *If from an even number an even number be subtracted, the remainder will be even.*

For from the even number *AB* let the even number *BC* be subtracted:
I say that the remainder *CA* is even [Fig. 2.5].

Fig. 2.5

For, since *AB* is even, it has a half part.   [VII. Def. 6]

For the same reason BC also has a half part;
so that the remainder [CA also has a half part, and] AC is therefore even.

Q.E.D.

PROPOSITION 25. *If from an even number an odd number be subtracted, the remainder will be odd.*

For from the even number AB let the odd number BC be subtracted;
I say that the remainder CA is odd [Fig. 2.6].

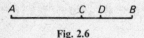

Fig. 2.6

For let the unit CD be subtracted from BC;
therefore DB is even. [VII. Def. 7]
But AB is also even;
therefore the remainder AD is also even. [IX. 24]
And CD is an unit;
therefore CA is odd. [VII. Def. 7]

Q.E.D.

PROPOSITION 26. *If from an odd number an odd number be subtracted, the remainder will be even.*

For from the odd number AB let the odd number BC be subtracted;
I say that the remainder CA is even [Fig. 2.7].

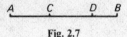

Fig. 2.7

For, since AB is odd, let the unit BD be subtracted;
therefore the remainder AD is even. [VII. Def. 7]
For the same reason CD is also even; [VII. Def. 7]
so that the remainder CA is also even. [IX. 24]

Q.E.D.

PROPOSITION 27. *If from an odd number an even number be subtracted, the remainder will be odd.*

For from the odd number AB let the even number BC be subtracted;
I say that the remainder CA is odd [Fig. 2.8].

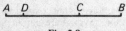

Fig. 2.8

Let the unit *AD* be subtracted;
therefore *DB* is even. [VII. Def. 7]
But *BC* is also even;
therefore the remainder *CD* is even. [IX. 24]
Therefore *CA* is odd. [VII. Def. 7]

Q.E.D.

PROPOSITION 28. *If an odd number by multiplying an even number make some number, the product will be even.**

For let the odd number *A* by multiplying the even number *B* make *C*;
I say that *C* is even [Fig. 2.9].

Fig. 2.9

For, since *A* by multiplying *B* has made *C*,
therefore *C* is made up of as many numbers equal to *B* as there are units
in *A*. [VII. Def. 15]
And *B* is even;
therefore *C* is made up of even numbers.
But, if as many even numbers as we please be added together, the whole
is even. [IX. 21]
Therefore *C* is even.

Q.E.D.

PROPOSITION 29. *If an odd number by multiplying an odd number make some number, the product will be odd.*

For let the odd number *A* by multiplying the odd number *B* make *C*;
I say that *C* is odd [Fig. 2.10].

For, since *A* by multiplying *B* has made *C*,

Fig. 2.10

---

* Book VII, Definition 15:
A number is said to **multiply** a number when that which is multiplied
is added to itself as many times as there are units in the other, and thus some
number is produced.

therefore C is made up of as many numbers equal to B as there are units in A. [VII. Def. 15]

And each of the numbers A, B is odd;

therefore C is made up of odd numbers the multitude of which is odd.

Thus C is odd. [IX. 23]

Q.E.D.

PROPOSITION 30. *If an odd number measure an even number, it will also measure the half of it.*

For let the odd number A measure the even number B;

I say that it will also measure the half of it [Fig. 2.11].

Fig. 2.11

For, since A measures B, let it measure it according to C;

I say that C is not odd.

For, if possible, let it be so.

Then, since A measures B according to C,

therefore A by multiplying C has made B.

Therefore B is made up of odd numbers the multitude of which is odd.

Therefore B is odd: [IX. 23]

which is absurd, for by hypothesis it is even.

Therefore C is not odd;

therefore C is even.

Thus A measures B an even number of times.

For this reason then it also measures the half of it.

Q.E.D.

The idea of a *figurate number* is completely foreign to the tradition of Euclid. It is, however, to be found in the *Introduction to Arithmetic* written by the neo-Pythagorean Nicomachus of Gerasa. Nicomachus was born in Judaea and seems to have lived in the second century AD. Theon of Smyrna, who made astronomical observations between AD 127 and 132, was also interested in figurate numbers and discusses them in his work on *The mathematical knowledge required for reading Plato.*

Nicomachus's work was translated into Latin by Apuleius of Madauros in the time of the Antonines. Boethius's adaptation of it, in his *De institutione arithmetica* written at the beginning of the sixth century AD, exerted great influence on Western scholars during the Middle Ages, and continued to be important as late as the sixteenth century.

Not all the complicated theory of figurate numbers can be traced back to Pythagoras, but there are good grounds for ascribing the basic ideas either to him or to his immediate followers. Points arranged in a straight line form a *linear number*. So every integer is linear. *Triangular numbers* are built up by placing one, two, three and more points on successive horizontal lines (Fig. 2.12).

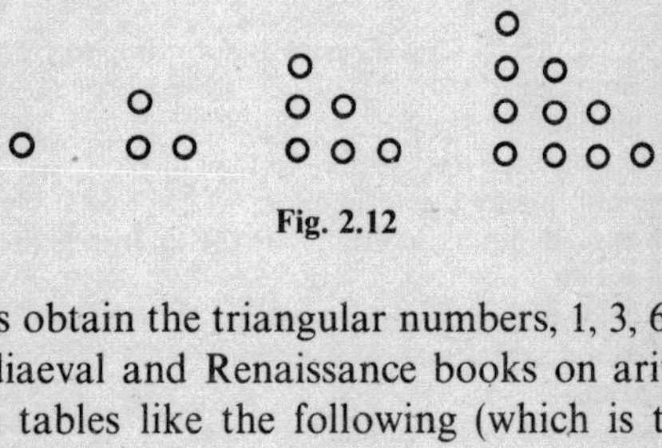

Fig. 2.12

We thus obtain the triangular numbers, 1, 3, 6, 10 etc.

In Mediaeval and Renaissance books on arithmetic we often find tables like the following (which is taken from Juan Martinez Siliceus's *Arithmetic*, printed in 1526):

Naturalis linea numerorum	1	2	3	4	5	6	7	8	9
Linea trigonalis	1	3	6	10	15	21	28	36	45

A *square number* is formed by points arranged in a square.

The Figure 2.13 shows the two possible ways of building up such a number: either by multiplying the side by itself or by adding up successive odd numbers. It thus demonstrates the formula $1 + 3 + 5 + \cdots + (2n - 1) = n^2$.

Fig. 2.13

An *oblong number* (ἑτερομήκης) can be arranged as a rectangle with one side one unit longer than the other (Fig. 2.14).

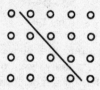

Fig. 2.14

It is clear from the figure that twice a triangular number is an oblong number. This is the equivalent of the formula

$$1 + 2 + 3 + 4 + \cdots + n = \frac{n(n + 1)}{2}.$$

Theon of Smyrna uses figurate numbers to show that the sum of two successive triangular numbers is a square, as can be seen in Fig. 2.15. There were also other types of

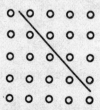

Fig. 2.15

plane numbers: *pentagonal numbers* (Fig. 2.16), which could be arranged as points defining a pentagon, namely the sums:

$$1, 1 + 4, 1 + 4 + 7, \quad \text{etc.,}$$

and also *hexagonal numbers*, *heptagonal numbers*, etc.

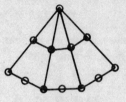

Fig. 2.16

After the plane figurate numbers come the solid ones. We shall be concerned only with pyramids and cubes. The latter are still familiar. The former are obtained by placing, one below the other in parallel planes, the first triangular number, 1, the second triangular number, 3, the third, 6, the fourth, 10, etc. (Fig. 2.17).

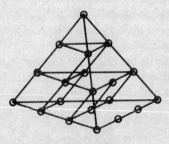

Fig. 2.17

The successive pyramidical numbers are therefore 1, 4, 10, 20, etc.

We do not know at what date the Greeks began to be interested in the theory of ratios and proportions, but we shall now discuss some arithmetical aspects of it. Such ideas appeared very early, probably among the Pythagorean school. In Book VII of Euclid's *Elements* we find as Definition 20:

Numbers are *proportional* when the first is the same multiple, or the same part, or the same parts, of the second that the third is of the fourth.

Nicomachus lists the various numerical ratios, and his nomenclature continued to be used until the seventeenth century. We shall discuss it briefly here.

Let us remember that our word ratio corresponds to the Greek word λόγος (logos) translated into Latin as 'proportio', and then, after 1500, as 'ratio'.

Nicomachus defines *multiple* ratios: e.g. double, triple, etc.; and their inverses or *submultiples*: sub-double, sub-triple, etc. Then follow the ratios ἐπιμόριος or superparticularis. In these ratios the first (antecedent) term contains the second (consequent) term and one further part of it.* We thus have the ratios ἡμιόλιος (sesqualter, $1\frac{1}{2}$) and ἐπίτριτος (sesquitertius, $1\frac{1}{3}$), etc. Their inverses are subsesquialter, subsesquitertius, etc. There is also the type of ratio called ἐπιμερής (superpartiens), in which the antecedent contains the consequent and several parts of it: e.g. $1\frac{2}{3}$ is ἐπιδιμερής (superbipartiens). Next comes the type called πολλαπλασιεπιμόριος (multiplex superparticularis) such as $3\frac{1}{2}$, and the type called πολλαπλασιεπιμερής (multiplex superpartiens) such as $2\frac{4}{5}$. The corresponding inverse ratio has the prefix ὑπο- (sub-). For example, we have quadruplex-superquadrupartiens giving a ratio for 5 to 24—the consequent contains the antecedent four and four fifths times.

Such clumsy terminology could only hamper mathematicians. It continued in use, however, and our illustration on page 14 shows a page of a manuscript by Nicholas Chuquet, dated 1484, which gives a diagram designed to

---

* i.e. $x = (1 + 1/n)y$ where n is a positive integer greater than 1.

represent this nomenclature. Chuquet's diagram is far from being the last of its kind.

The technique used to calculate proportions is much more significant.

We find an example in Book VII of Euclid:

PROPOSITION 13. *If four numbers be proportional, they will also be proportional alternately.*

That is if $a:b = c:d$ then $a:c = b:d$.

PROPOSITION 14. *If there be as many numbers as we please, and others equal to them in multitude, which taken two and two are in the same ratio, they will also be in the same ratio ex aequali.*

This means that if $a:b = d:e$ and $b:c = e:f$ then $a:c = d:f$. The ratio a to c is said to be *composed* from the ratios a to b and b to c. If, for example, $a:b = b:c$ the ratio $a:c$ is said to be *double* the other two. If $a:b = b:c = c:d$ the ratio $a:d$ is said to be *triple* the others.

Proportion by converse ratio or *invertendo*:

From $a:b = c:d$ we deduce $b:a = d:c$.

Proportion by composition of ratios or *componendo*:

From $a:b = c:d$ we deduce $(a + b):b = (c + d):d$.

Proportion by division of ratios or *dividendo*:

From $a:b = c:d$ we deduce $(a - b):b = (c - d):d$.

Proportion by conversion of ratios or *convertendo*:

From $a:b = c:d$ we deduce $a:(a - b) = c:(c - d)$.

This type of terminology, which fell into disuse only in the course of the seventeenth century, is used in all Greek mathematical work.

Another branch of arithmetic which has sometimes been traced back to the early Pythagoreans is the theory of medians or means, which is closely connected with the theory of proportions.

Archytas, a contemporary of Plato, defined the arithmetic mean in the following terms: An arithmetic mean is obtained when, of three numbers, the largest is greater than the second largest by the same amount that the second largest is greater than the smallest.

The arithmetic mean was much used by the Babylonians, so it is hardly surprising that Pythagoras should have been familiar with such a simple idea. Archytas goes on to define the geometric mean: The first number is to the second as the second is to the third:

$$\frac{a}{b} = \frac{b}{c}.*$$

Another type of mean, earlier called the 'sub-contrary mean', is called the 'harmonic mean' by Archytas who defines it by the relation that 'by what fraction of itself the first exceeds the second the second exceeds the third by that same fraction'. That is, if b is the harmonic mean of a and c, then

$$a:(a - b) = c:(b - c).$$

It is easy to show that $1/b$ is then the arithmetic mean of $1/a$ and $1/c$.

The Babylonians, who were equally ready to use numbers and their reciprocals, must therefore have known this mean. Their having done so would explain the origin of 'sub-contrary mean'. The name 'harmonic mean' is connected with the musical theories of the Pythagoreans.

---

* This is a case of *continued proportion*. The general case is what we should call a geometrical progression $a:b = b:c = c:d = d:e$.

Separated proportion occurs when the middle two terms are not the same.

Iamblichus reported that a pupil of Pythagoras, Aristaeus of Croton, said that the proportion

$$6:8 = 9:12$$

was taught to the Master 'by the Babylonians'. According to Nicomachus, Philolaus, a Pythagorean of the fourth century BC, referred to the cube as a *geometrical harmony*, because the number of its faces (6), its vertices (8) and its edges (12) defined a harmonic mean.

As for Pythagorean geometry we now know that Pythagoras's theorem was not discovered either by Pythagoras or by his followers. This fundamental theorem of metrical geometry was in fact known to the Babylonians.*

Speusippus, a follower of the Pythagoreans, is said to have written a book on 'the five figures associated with the elements of the Universe, their individual and relative properties'.

These five figures are the five regular polyhedra: the tetrahedron, the hexahedron or cube, the octahedron, the dodecahedron and the icosahedron.

Euclid does not define the regular tetrahedron in the *Elements* although in Book XIII he does give a detailed discussion of its properties as well as those of the other four regular polyhedra.

We find as Definition 12 at the beginning of Book XI:

A *pyramid* is a solid figure, contained by planes, which is constructed from one plane to one point.

Heron expands this in his own set of definitions, adding that a pyramid is said to be equilateral if it is circumscribed by four equilateral triangles. This figure is also called a tetrahedron.

Euclid, Book XI states [9]:

---

* See Chapter 4, Vol. 2.

DEFINITION 25. A *cube* is a solid figure contained by six equal squares.

DEFINITION 26. An *octahedron* is a solid figure contained by eight equal and equilateral triangles.

DEFINITION 27. An *icosahedron* is a solid figure contained by twenty equal and equilateral triangles.

DEFINITION 28. A *dodecahedron* is a solid figure contained by twelve equal, equilateral, and equiangular pentagons.

Some idea of the way Pythagoras and his followers regarded these regular polyhedra and how they 'associated them with the Elements of the Universe', may be obtained from the following passage from Plato's dialogue *Timaeus*. Timaeus, a legendary Pythagorean whom Plato perhaps regards as an idealized portrait of Archytas, explains his ideas to Socrates, Hermocrates and Critias:

In the first place it is clear to everyone that fire, earth, water, and air are bodies, and all bodies are solids. All solids again are bounded by surfaces, and all rectilinear surfaces are composed of triangles. There are two basic types of triangle, each having one right angle and two acute angles: in one of them these two angles are both half right angles, being subtended by equal sides, in the other they are unequal, being subtended by unequal sides. This we postulate as the origin of fire and the other bodies, our argument combining likelihood and necessity; their more ultimate origins are known to god and to men whom god loves. We must proceed to enquire what are the four most perfect possible bodies which, though unlike one another, are some of them capable of transformation into each other on resolution. If we can find the answer to this question we have the truth about the origin of earth and fire and the two mean terms between them; for we will never admit that there are more perfect visible bodies than these, each in its type. So we must do our best to construct four types of perfect body and maintain that we have grasped their nature sufficiently for our purpose. Of the two basic triangles, then, the isosceles has only one variety, the scalene an infinite number. We must therefore choose, if we are to start according to our own principles, the most perfect of this infinite number. If anyone can tell us of a better choice of triangle for the construction of the four bodies, his criticism will be welcome; but for our part we propose to pass over all the rest and pick on a single type, that of which a pair compose an equilateral triangle. It would be too long a story to give the reason, but if anyone can produce a proof that it is not so we will welcome his achievement. So let us assume that these are the two triangles from which fire and other bodies are constructed, one isosceles and the other having a greater side whose square is three times that of the lesser.* We must now proceed to clarify something we left undetermined a moment ago. It appeared as if all four types of body could pass into each other in the process of change; but this appearance is misleading. For, of the four bodies that are produced by our chosen types of triangle, three are composed of the scalene, but the fourth alone from the isosceles. Hence all four cannot pass into each other on resolution, with a large number of smaller constituents forming a lesser number of bigger bodies and vice versa; this can only happen with three of them. For these are all composed of one triangle, and when larger bodies are broken up a number of small bodies are formed of the same constituents, taking on their appropriate figures; and when small bodies are broken up into their component triangles a single new larger figure may be formed as they are unified into a single solid.

So much for their transformation into each other. We must next describe what geometrical figure each body has and what is the number of its components. We will begin with the construction of the simplest and smallest figure. Its basic unit is the triangle whose hypotenuse is twice the length of its shorter side. If two of these are put together with the hypotenuse as diameter of the resulting figure, and if the process is repeated three times and the diameters and shorter sides of the three figures are made to coincide in the same vertex, the result is a single equilateral triangle composed of six basic units [Fig. 2.18]. And if four equilateral triangles are put together, three of their plane angles meet to form a single solid angle, the one which comes next after the most obtuse of plane angles: and when four such angles have been formed the result is the simplest solid figure, which divides the surface of the sphere circumscribing it into equal and similar parts.

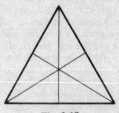

Fig. 2.18

The second figure is composed of the same basic triangles put together to form eight equilateral triangles, which yield a single solid angle from

---

* That is: the square of the larger side is three times that of the small one. See below, p. 58.

four planes. The formation of six such solid angles completes the second figure.

The third figure is put together from one hundred and twenty basic triangles, and has twelve solid angles, each bounded by five equilateral plane triangles, and twenty faces, each of which is an equilateral triangle.

After the production of these three figures of our basic units is dispensed with, and the isosceles triangle is used to produce the fourth body. Four such triangles are put together with their right angles meeting at a common vertex to form a square. Six squares fitted together complete eight solid angles, each composed by three plane right angles. The figure of the resulting body is the cube, having six plane square faces.

There still remained a fifth construction, which the god used for arranging the constellations on the whole heaven.* . . .

. . . Let us assign the cube to earth; for it is the most immobile of the four bodies and the most retentive of shape, and these are characteristics that must belong to the figure with the most stable faces. And of the basic triangles we have assumed, the isosceles has a naturally more stable base than the scalene, and of the equilateral figures composed of them the square is, in whole and in part, a firmer base than the equilateral triangle. So we maintain our principle of likelihood by assigning it to earth, while similarly we assign the least mobile of the other figures to water, the most mobile to fire, and the intermediate to air. And again we assign the smallest figure to fire, the largest to water, the intermediate to air; the sharpest to fire, the next sharpest to air, and the least sharp to water. So to sum up, the figure which has the fewest faces must in the nature of things be the most mobile, as well as the sharpest and most penetrating, and finally, being composed of the smallest number of similar parts, the lightest. Our second figure will be second in all these respects, our third will be third. Logic and likelihood thus both require us to regard the pyramid as the solid figure that is the basic unit or seed of fire; and we may regard the second of the figures we constructed as the basic unit of air, the third of water. We must, of course, think of the individual units of all four bodies as being far too small to be visible, and only becoming visible when massed together in large numbers. . . .

When earth meets fire it will be dissolved by its sharpness, and, whether dissolution takes place in fire itself or in a mass of air or water, will drift about until its parts meet, fit together and become earth again; for they can never be transformed into another figure. But when water is broken up by fire or again by air, its parts can combine to make one of fire and two of air; and the fragments of a single particle of air can make two of fire.†

---

* Plato means the dodecahedron.

† Compare Philolaus: There are five substances in the sphere: fire, water, earth and air inside the sphere, and the substance of the sphere itself making the fifth.

The five cosmic figures or Platonic solids (Plate III) seem to have been known from very ancient times. An Etruscan regular dodecahedron, found near Padua in 1885, apparently dates from the first half of the first millenium BC.*

The importance accorded to the equilateral triangle is indicated by the two following facts: that the first proposition in Euclid's *Elements* deals with constructing such a triangle, and that the angle of the equilateral triangle is divided into sixty degrees. This division into sixty goes back to the Babylonians, and seems to indicate that they had adopted this angle rather than the right angle as their unit. The decision must, however, have been taken, in connection with astronomical work, so it cannot date from earlier than 700 BC.

Many Babylonian tablets refer to the square, and we even find numerical information about equilateral triangles, squares, regular pentagons, and regular hexagons and heptagons.

However, the very important part played by the pentagon in Pythagorean mathematics is certainly connected with work on the icosahedron and the dodecahedron. The faces of the dodecahedron are pentagonal and pentagons occur in the icosahedron as the base of each of the pyramids formed by the faces which meet at a common vertex.

The Pythagoreans even adopted the five-pointed star or pentagram, the stellated pentagon, as their emblem which is thought to have mystic significance.

A pentagon can be inscribed in a circle using a straight edge and compasses.† It is possible that this discovery

---

* Gallo-Roman dodecahedra are to be found in several museums, such as the Antiquaires at Poitiers. Their faces contain circular holes of various diameters, and little spheres are attached to their vertices. Their use is not known.

† See Vol. 2, p. 127 for Ptolemy's elegant construction.

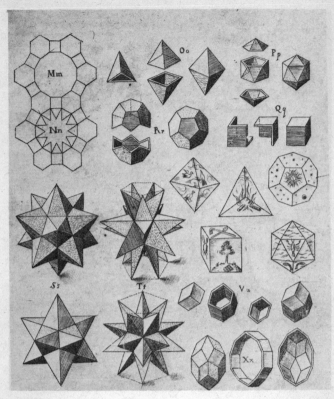

B.N.

PLATE III. The five cosmic figures illustrated in Kepler's *Harmonice Mundi* (1619).

was made by the early Pythagoreans, which would account for the school's interest in the figure.

To inscribe a heptagon in a circle, on the other hand, requires the trisection of an angle.

The early Pythagoreans are also credited with having discovered incommensurable (irrational) numbers. We

shall refer later (Vol. 2, p. 116) to a proof outlined by
Aristotle and developed in an appendix* to Book X of
Euclid, that the diagonal of a square is not commensurable
with the side of the square.

---

* X, 117, in texts, but now considered an interpolation and relegated to
an appendix. See Heath J.V.F.

3. Prelude

'Αγεωμέτρητος μηδεὶς εἰσίτω*
                Plato

We have traced the development of a philosophical type of mathematics which was to lead to the Neopythagoreanism and Neoplatonism of Late Antiquity, creating a tradition of mystical thought which moved further and further away from what we should now regard as Science.

Greek Mathematics was not, however, entirely of this kind. It was also bound up with the arts of sculpture, architecture, painting, the design of stage sets† and also music, which was much practised by the Pythagoreans. Moreover, it contributed to the techniques of engineering.

An important piece of construction work carried out on the Island of Samos about 530 BC by the engineer Eupalinos of Megara is described by Herodotus and was in fact uncovered by some German archaeologists in 1882. It is an aqueduct in the form of a tunnel one kilometre long, two metres wide and two metres high, with a channel running down the middle. The tunnel is almost straight, and vertical ventilation shafts are spaced out along it. Tunnelling was started from both ends and the two tunnels met in the middle with a horizontal error of 10 metres and a vertical error of 3 metres [1].

---

* Let no-one enter unless he is a geometer.

† I.e. scenery intended to give the illusion of three dimensions (applied perspective). J.V.F.

The geometrical technique used was probably the one later described by Heron of Alexandria in his treatise on the *Dioptra** (see Fig. 3.1).

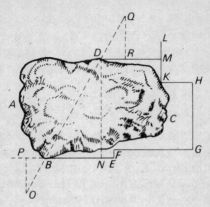

Fig. 3.1

If we want to drive a tunnel from B to D through a hill $ABCD$ we take sightings in two perpendicular directions DR and LK, so that by measuring the segments DM, MK, etc., we can calculate the lengths BN and DN, i.e. find two sides of the right-angled triangle BDN. At D and B we then construct the triangles DRQ and BPO, similar to the triangle BDN, and thus obtain the lines DQ and BO which indicate the direction in which the tunnel must go. This enables the tunnel to be started from both ends at once.

Eupalinos's work shows that by the end of the sixth century BC the Greeks were already capable of sophisticated practical geometry. A century later Hippocrates of Chios and Hippias of Elis were doing theoretical work in geometry. We shall discuss this work later, in Chapters 6 and 7 of Vol. 2.

---

* The dioptra was a sighting instrument, with pinholes, used in the manner of a modern theodolite.

At this point we shall concern ourselves mainly with the mathematicians of the fourth century BC, that is, of the time of Plato and Aristotle. We have already mentioned Archytas, a Pythagorean who studied geometry,* music and mechanics. The citizens of his native Tarentum made him their military leader. He was a friend of Plato's, and intervened on his behalf with Denys, ruler of Syracuse, to obtain permission for the philosopher to leave Sicily and return to Athens.

Prominent among Plato's friends there is also another figure, perhaps as purely mythical as Timaeus of Locri, namely Theaetetus. All we know about him is derived from Plato, in fact from the dialogue *Theaetetus or On the subject of Science'*, which begins with a conversation between Euclid of Megara and Terpsion:†

> *Euclides.* Have you only just arrived from the country, Terpsion?
> *Terpsion.* No, I came some time ago: and I have been in the Agora looking for you, and wondering that I could not find you.
> *Euc.* But I was not in the city.
> *Terp.* Where then?
> *Euc.* As I was going down to the harbour, I met Theaetetus—he was being carried up to Athens from the army at Corinth.
> *Terp.* Was he alive of dead?
> *Euc.* He was scarcely alive, for he has been badly wounded; but he was suffering even more from the sickness which has broken out in the army.
> *Terp.* The dysentery, you mean?
> *Euc.* Yes.
> *Terp.* Alas! what a loss he will be!
> *Euc.* Yes, Terpsion, he is a noble fellow; only today I heard some people highly praising his behaviour in this very battle.
> *Terp.* No wonder; I should rather be surprised at hearing anything else of him. But why did he go on, instead of stopping at Megara?
> *Euc.* He wanted to get home: although I entreated and advised him to remain, he would not listen to me; so I set him on his way, and turned back, and then I remembered what Socrates had said of him, and thought how remarkably this, like all his predictions, had been fulfilled. I believe that he had seen him a little before his own death, when Theaetetus was a youth, and he had a memorable conversation with him, which he repeated

---

* See Vol. 2, p. 161 for his solution to the problem of doubling the cube.
† We quote only part of it. The translation is by Jowett.

to me when I came to Athens; he was full of admiration of his genius, and
said that he would most certainly be a great man, if he lived.

Euclid had noted down Socrates's recollections of this
meeting, and at Terpsion's request he has them read out
by a slave. The discussion turns to mathematics:

Theaet.... You mean, if I am not mistaken, something like what
occurred to me and to my friend, here, your namesake Socrates, in a recent
discussion.

Soc. What was that, Theaetetus?

Theaet. Theodorus* was writing out for us something about roots, such
as the sides of squares three or five feet in area† showing that they are
incommensurable by the unit: he took the other examples up to seventeen,
but there for some reason he stopped. Now as there are innumerable such
roots, the notion occurred to us of attempting to find some common
description which can be applied to them all.

Soc. And did you find any such thing?

Theaet. I think that we did; but I should like to have your opinion.

Soc. Let me hear.

Theaet. We divided all numbers into two classes: those which are made
up of equal factors multiplying into one another, which we compared to
square figures and called square or equilateral numbers;—that was one
class.

Soc. Very good.

Theaet. The intermediate numbers, such as three and five, and every other
number which is made up of unequal factors, either of a greater multiplied
by the less, or of a less multiplied by a greater, and, when regarded as a
figure, is contained in unequal sides;—all these we compared to oblong
figures, and called them oblong numbers.

Soc. Capital; and what followed?

Theaet. The lines, or sides, which have for their squares the equilateral
plane numbers, were called by us lengths; and the lines whose squares are
equal to the oblong numbers, were called powers or roots; the reason of
this latter name being, that they are commensurable with the former not in
linear measurement, but in the area of their squares. And a similar distinction was made among solids.

We must not expect mathematical rigour from such a text.
It does, however, show that at the time when Plato wrote

---

* Theaetetus and the young Socrates were both pupils of this teacher.
He was traditionally supposed to have taught Plato mathematics, when
Plato stayed with him in Cyrene, on his way to Egypt. He is known to us
only as one of the characters in the Dialogue *Theaetetus*.

† Squares with sides of 3 or 5 feet.

this dialogue, about 367 or 384 BC, the theories which were eventually to find expression in Book X of Euclid's *Elements** were already taking shape.

We now come to two very important mathematicians, Democritus of Abdera and Eudoxus of Cnidus.

Democritus seems to have lived from about 460 until about 370 BC. He was interested in all branches of the philosophy and science of his time.† In the first four centuries AD many alchemical writings were attributed to Democritus. Some can be dated to as far back as the third century BC, but by then Callimachus had already pointed out that they were not by Democritus.

Eudoxus lived in the first half of the fourth century BC. He practised as a doctor but was famous mainly as an astronomer and as a mathematician.

Archimedes mentions these two scholars in his introduction to his treatise *The Method*:

This is a reason why in the case of the theorems the proof of which Eudoxus was the first to discover, namely that the cone is a third part of the cylinder, and the pyramid of the prism, having the same base and equal height, we should give no small share of the credit to Democritus who was the first to make the assertion with regard to the said figures though he did not prove it.

The two propositions mentioned by Archimedes are proved in Book XII of Euclid's *Elements*, and we have reason to believe that all the material of Book XII was, at least in outline, the work of Eudoxus and his school.

Here we shall consider only the volume of the pyramid. We know from the Moscow Papyrus that the Egyptians had a method of calculating the volume of a frustum of a pyramid which was equivalent to using the formula:

$$V = \frac{h}{3}(a^2 + ab + b^2).$$

---

* See Vol. 2, pp. 95–104.

† This is the Democritus who originated the theory of Atomism.

This formula can be derived very easily in the special case where we have a rectangular parallelepiped (see Fig. 3.2).

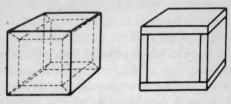

Fig. 3.2

Let us imagine a box in the shape of a cube, with external edge *a* and internal edge *b* (*left-hand figure*). The panels are of thickness *h*. Each of the six panels can be considered as a frustum of a pyramid with thickness *h* and bases of areas a^2 and b^2, and of volume *V*. The total volume of the panels is therefore 6*V*. The same box could be made more crudely from six panels each in the form of a parallelepiped, each with thickness *h* (*right-hand figure*). Two would be square, with side *a*, two would be rectangular with sides *a* and *b*, and two square with side *b*. The total volume, 6*V*, is therefore

$$2a^2h + 2abh + 2b^2h,$$

and we have obtained the Egyptian formula.

Such intuitive reasoning is common. An instance of it is provided by a passage from Leclerc's *Treatise on Geometry* (*Traité de Géométrie*), published in 1694:

I shall prove that the volume of a pyramid is found by multiplying a third of its height by its base.

Let us suppose that the six faces of a cube *HB* are the bases of six pyramids, each having its vertex at the centre *A*. These six pyramids which go to make up the cube will be equal [Fig. 3.3, *left-hand figure*].

If the side *BC* is of length 12 inches the base *BCDE* will have an area of 144 square inches, and the whole cube will have a volume of 1728 cubic inches, a sixth of which, 288 cubic inches, will go to make up the volume of each pyramid.

But since the entire cube is 12 inches in height, the height of the pyramid *ABCDE* will be 6 inches, and one third of 6, multiplied by the base *BCDE*, that is 2 times 144, will give the same result of 288 cubic inches which we found to be the volume of each pyramid. So the volume of a pyramid can be found by multiplying the base by one third of the height [Fig. 3.3].

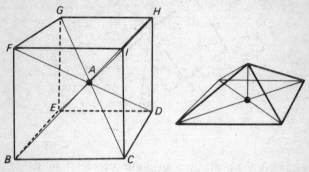

Fig. 3.3

There is no way of knowing whether Democritus learned how to find the volume of a pyramid from the Egyptians or whether he worked it out independently. In any case, Archimedes states quite clearly that Democritus was the first Greek to describe this method in public. The intuitive procedures we have described *suggest* how the volume can be calculated, but do not establish a general rule. The next step is to be found in Book XII of Euclid's *Elements*. Proposition 3 reads:

Any pyramid which has a triangular base is divided into two pyramids equal and similar to one another, similar to the whole and having triangular bases, and into two equal prisms; and the two prisms are greater than the half of the whole pyramid [2].

Consider the triangular pyramid *ABCD* in figure 3.4. The points *E*, *F*, *G*, *H*, *I* and *J* are the mid-points of its edges. The two pyramids, which are stated to be 'equal and similar' (today we should merely call them equal), are *AEFG* and *FHCI*. The two 'equal' (equivalent) prisms are *BHJEFG* and *HFIJGD*.

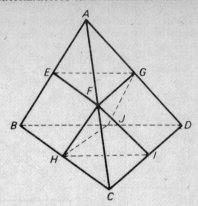

Fig. 3.4

Each of the prisms has a volume one eighth of that of the prism with base BCD and edge BA, so we can deduce that the volume of the pyramid is between one quarter and one half of the volume of the prism. An Egyptian mathematician, dealing only with fractions which were the reciprocal of an integer, would deduce that the correct ratio was one third.

The idea of similarity is an intuitive one which is found, implicitly, throughout Egyptian and Babylonian mathematical work: if two figures have properties, volumes for example, in the ratio r, and we copy these figures on the same scale, then the copies will have properties in the same ratio r. Suppose we are required to find the ratio between the volume of the given pyramid and the volume of the prism on the same base BCD and with the same vertex A. Let this ratio be x.

The pyramid $AEFG$ and the prism $BHJEFG$ are half-scale copies of the original pyramid and prism. So if the volume of the small prism is 1, that of the small pyramid will be x.

But the complete pyramid contains 2 equivalent prisms and 2 equal pyramids. Its volume is therefore $2 + 2x$.

Moreover, the large prism has volume 8, so the large pyramid has volume $8x$, and we can deduce that

$$8x = 2 + 2x$$

i.e. $\quad x = \frac{1}{3}$.

Perhaps Democritus had reached this stage, where the reasoning is already abstract but not such as would have satisfied the geometers of the third century BC. Archimedes would not have felt justified in calling this a proof.

Eudoxus abandons the intuitive idea of similarity and uses an argument which involves infinitesimals. In the second part of the last proposition he notes that each of the little pyramids is of smaller volume than each of the prisms, and deduces that the sum of the two prisms is greater than half the complete pyramid.

But if from some value one subtracts more than half, and then subtracts from the remainder more than half of that remainder, and so on, one ends up with a remainder smaller than any given quantity [Aristotle, and Euclid Book X].

This line of argument is followed by Euclid in Book XII, and he presumably took its outlines from Eudoxus:

PROPOSITION 4. *If there be two pyramids of the same height which have triangular bases, and each of them be divided into two pyramids equal to one another and similar to the whole, and into two equal prisms, then, as the base of the one pyramid is to the base of the other pyramid, so will all the prisms in the one pyramid be to all the prisms, being equal in multitude, in the other pyramid* [3].

PROPOSITION 5. *Pyramids which are of the same height and have triangular bases are to one another as the bases* [4].

The argument goes:
Let P_1 and P_2 be the volumes of the pyramids and B_1 and B_2 their bases. We wish to show that $B_1 : B_2 = P_1 : P_2$.
Suppose the proposition to be untrue, and let X be a volume such that $B_1 : B_2 = P_1 : X$. Now X is either smaller or larger than P_2.

Suppose first of all that X is smaller than P_2. We use the method of the previous two theorems to divide the two pyramids into the same number of prisms and pyramids, until the difference between P_2 and S_2, the sum of the prisms inscribed in the second pyramid, is less than $P_2 - X$. So $X < S_2 < P_2$.

If S_1 is the sum of the prisms inscribed in the first pyramid $S_1 < P_1$.

Now the proposition states that

$$B_1 : B_2 = S_1 : S_2.$$

But

$$B_1 : B_2 = P_1 : X,$$

so

$$P_1 : X = S_1 : S_2$$

and by permutation

$$P_1 : S_1 = X : S_2$$

which is absurd, because the ratio on the left is greater than unity (since $P_1 > S_1$) and the ratio on the right is less than unity (since $X < S_2$).

If X were greater than P_2 we should take $Y : P_2 = P_1 : X$ and Y would be less than P_2 because P_1 is less than X.

Now

$$B_1 : B_2 = P_1 : X = Y : P_2$$

so

$$B_2 : B_1 = P_2 : Y$$

which reduces to the previous case, with the pyramids exchanging rôles.

PROPOSITION 6. *Pyramids which are of the same height and have polygonal bases are to one another as the bases* [5].

This proposition reduces to the previous one if we divide the polygons up into triangles.

PROPOSITION 7. *Any prism which has a triangular base is divided into three pyramids equal to one another which have triangular bases* [6].

These would now be called 'equivalent pyramids'. Euclid's is still considered the classic proof.

The modern proof of Proposition 5 by subdividing the pyramid by means of planes parallel to the base, was first used in the seventeenth century, in the *Elements of Geometry* (*Eléments de Géométrie*) of Father Tacquet (1654), a work which paraphrases Euclid's *Elements*. Tacquet was influenced by Cavalieri's use of indivisibles and, even more, by the methods Archimedes employed in dealing with more complicated bodies.*

The study of the pyramid serves as an example to show how much ground was covered between Babylonian and Egyptian mathematics and the Greek geometry of the middle of the fourth century BC, which foreshadowed the great achievements of the Hellenistic period.

---

* We might be tempted to try to find the volume of the pyramid without using infinitesimals. The task is, in fact, impossible, as we know from a memoir published by Raoul Bricard in the *Nouvelles Annales de Mathématiques* in 1896.

4. The Lawgivers of Geometry*

*Qui Archimedem et Apollonium intellegit,
recentiorum summorum vivorum inventa
parcius mirabitur.†*

Leibniz

We now turn our attention to one of the greatest ages of
mathematics, the third century BC. We must begin by
acknowledging that we have very little exact chronological
data. For the three greatest masters, Euclid, Archimedes
and Apollonius, we can give only one definite date: that of
the death of Archimedes during the sack of Syracuse in the
Second Punic War, in the year 212 BC.

We know that Archimedes lived before Apollonius,
because he discusses conics in less sophisticated language.
Also, Apollonius in his prefaces speaks of Conon as one of
his predecessors, whereas Archimedes was a contemporary
of Conon and several times refers with regret to his 'recent'
death. It is thus quite easy to establish the relation between
Archimedes and Apollonius, but, although the latter lived
later than Euclid, we do not as yet have any means of
comparing the dates of Euclid and Archimedes.

---

* We have taken this title from Gino Loria, *Storia delle matematiche*,
2nd edition, Milan 1950.

† One who understands the work of Archimedes and Apollonius will
wonder little at the greatest discoveries of modern scholars.

For want of more reliable sources, we shall draw on tradition in what we have to say about each of these three great mathematicians.

4.1 Euclid

We have seen that at the beginning of 'Theaetetus', Plato introduces a philosopher called Euclid of Megara.

Western scholars long confused him with Euclid the mathematician. The confusion goes back at least to Valerius Maximus (second century) who, writing about doubling the cube, says that Plato sent the delegates from Delos to 'Euclid the Geometer'. An edition of the *Elements* printed in Basel in 1558 bears the title:

'Euclidis Megarensis mathematici clarissimi Elementorum geometricorum libri XV.'

The identification was considered erroneous by Constantine Lascaris, a Greek scholar who died in Messina in 1493, and in 1557, Stephanus Gracilis, a professor at Paris, stated in his preface to the Graeco-Latin edition of the propositions of the *Elements* that he shared Lascaris's opinion.

Lascaris cited as evidence a passage in Proclus's commentary on the first book of the *Elements*.

After referring to several fourth-century mathematicians, the last one being Philip of Mende, Proclus wrote [1]:

Not long after these men came Euclid, who brought together the *Elements*, systematizing many of the theorems of Eudoxus, perfecting many of those of Theaetetus, and putting in irrefutable demonstrable form propositions that had been rather loosely established by his predecessors. He lived in the time of Ptolemy the First, for Archimedes, who lived after the time of the first Ptolemy, mentions Euclid.* It is also reported that Ptolemy once asked Euclid if there was not a shorter road to geometry than through the *Elements*, and Euclid replied that there was no royal road

* There is no known passage where Archimedes mentions Euclid, unless we accept the clearly apocryphal passage in the Heiberg edition. *On the Sphere and Cylinder*, Book I, Proposition 2.

to geometry.* He was therefore later than Plato's group but earlier than Eratosthenes and Archimedes, for these two men were contemporaries, as Eratosthenes somewhere says. Euclid belonged to the persuasion of Plato and was at home in this philosophy† and this is why he thought the goal of the *Elements* as a whole to be the construction of the so-called Platonic figures.

Arabic tradition adds much to what we are told by Proclus. It relates that Euclid was the son of Naucrates and the grandson of Zenarchus or of Berenice; that he was of Greek descent; that he lived in Damascus although he had been born in Tyre; and that he was deeply versed in geometry. Why should we not believe this? . . . then, on the other hand, why should we? . . .

Pappus in the preface to Book VII of his *Mathematical Collection* tells us that Apollonius 'had long spent his leisure time with the pupils of Euclid in Alexandria, and had learned from them an attitude of mind not lacking in the wisdom of experience'. However, Hultsch, the most recent editor of Pappus, has doubts about the authenticity of this passage.

It is on such meagre testimony that we must base all our supposed knowledge of Euclid's life.

His works, fortunately, survive in plenty: thirteen books of the *Elements*, the *Data*, the *Phenomena*, the *Divisions* and the *Optics* have all come down to us. A *Catoptrics*, an *Introduction to Harmony* and a fragment called *Of light and heavy bodies* have survived and have been ascribed to Euclid, but are probably not by him.

We have an Arabic translation of his treatise on the *Division of Figures* and Pappus mentions *three books of Porisms, Loci on Surfaces* and *four Books on Conics*. It is to

---

* Stobaeus reports a similar conversation between the mathematician Menaechmus and Alexander the Great.

† This statement by Proclus, himself a Neo-platonist, is not supported by any other authority.

this last work that Apollonius refers in the preface to Book I of his *Conics*.

We shall concern ourselves only with the *Elements*.

BOOK I. The book begins with twenty three definitions, the most important being:*

1. A **point** is that which has no part.
2. A **line** is breadthless length.
3. The extremities of a line are points.
4. A **straight line** is a line which lies evenly with the points on itself.
5. A **surface** is that which has length and breadth only.
6. The extremities of a surface are lines.
7. A **plane surface** is a surface which lies evenly with the straight lines on itself.
8. A **plane angle** is the inclination to one another of two lines in a plane which meet one another and do not lie in a straight line.
9. And when the lines containing the angle are straight, the angle is called **rectilineal**.
10. When a straight line set up on a straight line makes the adjacent angles equal to one another, each of the equal angles is **right**, and the straight line standing on the other is called a **perpendicular** to that on which it stands.
13. A **boundary** is that which is an extremity of anything.
14. A **figure** is that which is contained by any boundary or boundaries.
15. A **circle** is a plane figure contained by one line such that all the straight lines falling upon it from one point among those lying within the figure are equal to one another.
16. And the point is called the **centre** of the circle.
17. A **diameter** of the circle is any straight line drawn through the centre and terminated in both directions by the circumference of the circle, and such a straight line also bisects the circle.
23. **Parallel** straight lines are straight lines which, being in the same plane and being produced indefinitely in both directions, do not meet one another in either direction.

Then come the *Demands*, Αἰτήματα. In the thirteenth century Latin translation by Campanus, the word is rendered as *Petitiones*. Zamberti's translation, at the beginning of the sixteenth century, gives 'Postulata' from which we derive the term 'Postulates'.

---

* The translations we give will generally be those of Heath, *The Thirteen Books of Euclid's Elements*.

Let the following be postulated:

1. To draw a straight line from any point to any point.
2. To produce a finite straight line continuously in a straight line.
3. To describe a circle with any centre and distance.
4. That all right angles are equal to one another.
5. That, if a straight line falling on two straight lines make the interior angles on the same side less than two right angles, the two straight lines, if produced indefinitely, meet on that side on which are the angles less than the two right angles.

After the definitions and the postulates Euclid states his *Common Notions*, Κοιναὶ ἔννοιαι.*

Their number varies from edition to edition. Only five are regarded as authentic in the light of modern criticism:

1. Things which are equal to the same thing are also equal to one another.
2. If equals be added to equals, the wholes are equal.
3. If equals be subtracted from equals, the remainders are equal.
[7] 4. Things which coincide with one another are equal to one another.
[8] 5. The whole is greater than the part.

In modern language the first three axioms assert:

1. If $A = B$ and if $A = C$, then $B = C$.
2. If $A = B$ and if $C = D$ then $A + C = B + D$.
3. If $A = B$ and if $C = D$ then $A - C = B - D$.

The axioms are explained in this algebraic manner in Hérigone's *Euclid* of 1639. We show a page of this work in Plate IV, opposite. It should be noted that the symbol used for equality is 2|2.

The fifth axiom could be expressed as

$$A + B > A.$$

---

* Campanus translates 'Communes animi conceptiones' and Zamberti 'Communes sententiae', while Stephanus Gracilis gives 'Communes notiones'. Clavius introduces the word 'axiom', borrowed from the usage of Aristotle, and is, in fact, very verbose: 'Communes notiones, sive Axiomata, quae et Pronunciata dici solent, vel dignitates.'

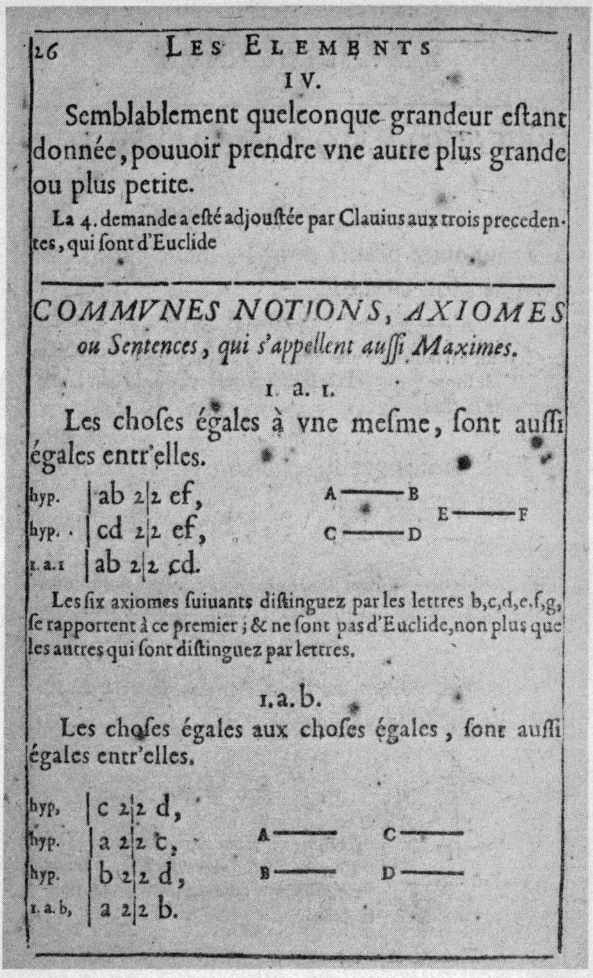

26   LES ELEMENTS
IV.

Semblablement quelconque grandeur estant donnée, pouuoir prendre vne autre plus grande ou plus petite.

La 4. demande a esté adjoustée par Clauius aux trois precedentes, qui sont d'Euclide

---

COMMVNES NOTIONS, AXIOMES
ou Sentences, qui s'appellent aussi Maximes.

I. a. I.

Les choses égales à vne mesme, sont aussi égales entr'elles.

hyp.	ab 2|2 ef,
hyp.	cd 2|2 ef,
I. a. I	ab 2|2 cd.

A——B
C——D   E——F

Les six axiomes suiuants distinguez par les lettres b,c,d,e,f,g, se rapportent à ce premier ; & ne sont pas d'Euclide, non plus que les autres qui sont distinguez par lettres.

I. a. b.

Les choses égales aux choses égales, sont aussi égales entr'elles.

hyp.	c 2|2 d,
hyp.	a 2|2 c,
hyp.	b 2|2 d,
I. a. b.	a 2|2 b.

A—— C——
B—— D——

B.N.

PLATE IV. A page from Hérigone's *Euclid* (1639).

The fourth is specifically geometrical.*

After this introduction the first book gives forty-eight propositions which can be divided into three groups [2].

FIRST GROUP. Twenty-six propositions referring to

(a) the construction of triangles;

(b) the relations between the elements of a triangle: sides and angles;

(c) three cases of equal triangles;

(d) some geometrical constructions with straight edge and compasses: bisecting an angle, finding the centre of a segment, constructing the perpendicular to a straight line.

PROPOSITION[†] 1. *On a given finite straight line to construct an equilateral triangle.*

PROPOSITION 2. *To place at a given point (as an extremity) a straight line equal to a given straight line.*

Let A be the given point, and BC the given straight line.

Thus it is required to place at the point A (as an extremity) a straight line equal to the given straight line BC [see Fig. 4.1].

From the point A to the point B let the straight line AB be joined; and on it let the equilateral triangle DAB be constructed.

Let the straight lines AE, BF be produced in a straight line with DA, DB; with centre B and distance BC let the circle CGH be described; and again, with centre D and distance DG let the circle GKL be described.

Then, since the point B is the centre of the circle CGH,

$$BC \text{ is equal to } BG.$$

Again, since the point D is the centre of the circle GKL,

$$DL \text{ is equal to } DG.$$

---

* We must distinguish between Euclid's 'equality' and the modern 'congruence'.

For Euclid congruence implies equality but equality does not imply congruence. Two rectangles $4\,m \times 3\,m$ and $2\,m \times 6\,m$ are equal in Euclid's sense. Today we should say they are equivalent, i.e. have the same area.

† The problems are distinguished from the theorems only by their final words 'which it was required to do' instead of 'which it was required to prove'.

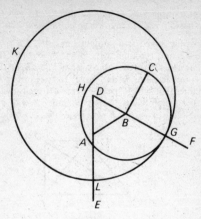

Fig. 4.1

And in these *DA* is equal to *DB*; therefore the remainder *AL* is equal to the remainder *BG*.

But *BC* was also proved equal to *BG*; therefore each of the straight lines *AL*, *BC* is equal to *BG*.

And things which are equal to the same thing are also equal to one another; therefore *AL* is also equal to *BC*.

Therefore at the given point *A* the straight line *AL* is placed equal to the given straight line *BC*. (Being) what it was required to do.

PROPOSITION 3. *Given two unequal straight lines, to cut off from the greater a straight line equal to the less.*

PROPOSITION 4. *If two triangles have the two sides equal to two sides respectively, and have the angles contained by the equal straight lines equal, they will also have the base equal to the base, the triangle will be equal to the triangle, and the remaining angles will be equal to the remaining angles respectively, namely those which the equal sides subtend.*

The proofs, as in modern elementary textbooks, involve transposition and proving that triangles are congruent.

PROPOSITIONS 5 and 6. *The angles at the base of an isosceles triangle are equal, and conversely.*

PROPOSITION 7. *Given two straight lines constructed on a straight line (from its extremities) and meeting in a point, there cannot be constructed on the same straight line (from its extremities), and on the same side of it, two other straight lines meeting in another point and equal to the former two respectively, namely, each to that which has the same extremity with it.*

For, if possible, given two straight lines *AC*, *CB* constructed on the straight line *AB* and meeting at the point *C*, let two other straight lines *AD*, *DB* be constructed on the same straight line *AB*, on the same side of it, meeting in another point *D* and equal to the former two respectively, namely, each to that which has the same extremity with it, so that *CA* is equal to *DA* which has the same extremity *A* with it, and *CB* to *DB* which has the same extremity *B* with it; and let *CD* be joined [Fig. 4.2].

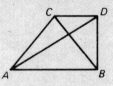

Fig. 4.2

Then, since *AC* is equal to *AD*, the angle *ACD* is also equal to the angle *ACD*; [I.5] therefore the angle *ACD* is greater than the angle *DCB*; therefore the angle *CDB* is much greater than the angle *DCB*. Again, since *CB* is equal to *DB*, the angle *CDB* is also equal to the angle *DCB*. But it was also proved much greater than it: which is impossible. Therefore, etc.

Q.E.D.

PROPOSITION 8. *If two triangles have two sides equal to two sides respectively, and have also the base equal to the base, they will also have the angles equal which are contained by the equal straight lines.*

One of the two triangles is moved so that its base is brought into contact with that of the other triangle. The previous theorem is then applied.

PROPOSITION 9. *To bisect a given rectilineal angle.*

PROPOSITION 10. *To bisect a given finite straight line.*

PROPOSITION 11. *To draw a straight line at right angles to a given straight line from a given point on it.*

The two constructions use the equilateral triangle.

PROPOSITION 12. *To a given infinite straight line, from a given point which is not on it, to draw a perpendicular straight line.*

PROPOSITION 13. *If a straight line set up on a straight line makes angles, it will make either two right angles or angles equal to two right angles.*

PROPOSITION 14. Converse of proposition 13.

PROPOSITION 15. Equality of opposite angles at a vertex.

PROPOSITION 16. In any triangle an exterior angle is greater than a non-adjacent interior angle.

PROPOSITION 17. *In any triangle two angles taken together in any manner are less than two right angles.*

PROPOSITIONS 18 and 19. Comparison of the angles and opposite sides of a triangle.

PROPOSITION 20. *In any triangle two sides taken together in any manner are greater than the remaining one.*

PROPOSITION 21. *If on one of the sides of a triangle, from its extremities, there be constructed two straight lines meeting within the triangle, the straight lines so constructed will be less than the remaining two sides of the triangle, but will contain a greater angle.*

PROPOSITION 22. Construction of a triangle, given the three sides; and the conditions for the construction to be possible.

PROPOSITION 23. *On a given straight line and at a point on it to construct a rectilineal angle equal to a given rectilineal angle.*

We go back to the preceding proposition.

PROPOSITION 24. *If two triangles have two sides equal to two sides respectively, but have one of the angles contained by the equal straight lines greater than the other, they will also have the base greater than the base.*

PROPOSITION 25. Converse of the preceding one.

PROPOSITION 26. *If two triangles have the two angles equal to two angles respectively, and one side equal to one side, namely, either the side adjoining the equal angles, or that subtending one of the equal angles, they will also have the remaining sides equal to the remaining sides and the remaining angle to the remaining angle.*

Proof by *reductio ad absurdum*, not involving transposition, using propositions 4 and 16.

SECOND GROUP. This includes Propositions 27 to 32.
It deals with parallel lines and using them to deduce the
sum of the angles of a triangle.

PROPOSITION 27. *If a straight line falling on two straight lines make the
alternate angles equal to one another, the straight lines will be parallel to one
another.*

Proof by *reductio ad absurdum.*

PROPOSITION 28. *If a straight line falling on two straight lines make the
exterior angle equal to the interior and opposite angle on the same side, or
the interior angles on the same side equal to two right angles, the straight
lines will be parallel to one another.*

PROPOSITION 29. *A straight line falling on parallel straight lines makes the
alternate angles equal to one another, the exterior angle equal to the interior
and opposite angle, and the interior angles on the same side equal to two right
angles.*

It is here that we have the first appeal to Demand 5,
Euclid's Postulate (p. 70). We may note that since he leaves
the use of this postulate until as late as possible Euclid
could be considered as the first 'non-Euclidean' geometer.

PROPOSITION 30. *Straight lines parallel to the same straight line are also
parallel to each other.*

PROPOSITION 31. *Through a given point to draw a straight line parallel
to a given straight line.*

Solution by construction of equal alternate angles.

PROPOSITION 32. *In any triangle, if one of the sides be produced, the
exterior angle is equal to the two interior and opposite angles, and the three
interior angles of the triangle are equal to two right angles.*

THIRD GROUP. The third group of Propositions, from
Proposition 33 to Proposition 48, is mainly concerned with
the equivalence of polygons, here divided up into parallelograms and triangles. Propositions 33 and 34 start by
explaining the concept of a parallelogram.

PROPOSITION 33. *The straight lines joining equal and parallel straight lines (at the extremities which are) in the same directions (respectively) are themselves also equal and parallel.*

PROPOSITION 34. *In parallelogrammic areas the opposite sides and angles are equal to one another, and the diameter bisects the areas.**

The proposition allows us to add the parallelogram to the four figures mentioned in Definition 22 (below).

DEFINITION 22. Of quadrilateral figures, a **square** is that which is both equilateral and right-angled; an **oblong** that which is right-angled but not equilateral; a **rhombus** that which is equilateral but not right-angled; and a **rhomboid** that which has its opposite sides and angles equal to one another but is neither equilateral nor right-angled. And let quadrilaterals other than these be called **trapezia**.

PROPOSITION 35. *Parallelograms which are on the same base and in the same parallels are equal to one another.*

'Equal' here means 'equivalent'.

PROPOSITION 36. *Parallelograms which are on equal bases and on the same parallels are equal to one another.*

PROPOSITION 37. *Triangles which are on the same base and in the same parallels are equal to one another.*

PROPOSITION 38. *Triangles which are on equal bases and in the same parallels are equal to one another.*

PROPOSITION 39. *Equal triangles which are on the same base and on the same side are also in the same parallels.*

PROPOSITION 41. *If a parallelogram has the same base with a triangle and be in the same parallels, the parallelogram is double of the triangle.*

PROPOSITION 42. *To construct, in a given rectilineal angle, a parallelogram equal to a given triangle.*

---

* In the following propositions Euclid no longer speaks of parallelogrammic areas but of parallelograms.

Note the word 'diameter' where we should say 'diagonal'. Both words are Greek, but the second only appears once in the *Elements*, in Book XI, Proposition 28. This passage could have been modified by later editors, such as Theon of Alexandria. Archimedes always used 'diameter'. 'Diagonal' is to be found in the *Geometrical Definitions* attributed to Heron.

PROPOSITION 43. *In any parallelogram the complements of the parallelograms about the diameter are equal to one another.*

Let *ABCD* be a parallelogram [Fig. 4.3], and *AC* its diameter; and about *AC* let *EH*, *FG* be parallelograms, and *BK*, *KD* the so-called complements; I say that the complement *BK* is equal to the complement *KD*.

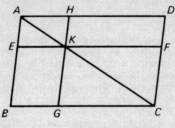

Fig. 4.3

In his *Practical Geometry* (*Géométrie Practique*) of 1702 Manesson Mallet illustrated this proposition with the following anecdote (see Plate V):

'A Referendary, observing that his garden, shown in the figure as the rectangle *MNOP*, was too short, asked his brother-in-law, who was secretary to the King and had studied Geometry, how far beyond the line *RO* the garden would extend if he gave the part *MQRP*, his maze, to the owner of the tile factory *POST* which adjoined his garden along the wall *PO*.

'The brother-in-law solved the problem as follows: he had a cord stretched along *QR*, the side of the Maze, and passed it into the tile factory through a hole *R* in the wall *PO*. This cord is shown in the diagram as *QRY*. He then went to the far corner of the garden, *N*, and had a cord stretched from there via the corner *QRP* as far as the point *V* where it met the wall *PT*. We show this cord as *NRV*. At *V* he drove in a stake, and from this stake, ran a cord *VX* across the tile factory, parallel to the wall *PO*, until it met the wall *OS*.

'He then noted the position of the point Z where the cord VX crossed the cord QRY, and told his brother-in-law that the rectangle $ROXZ$ was the piece of garden he would gain if he gave away $MQRP$, the area of his Maze. He thus answered the question that had been put to him.'

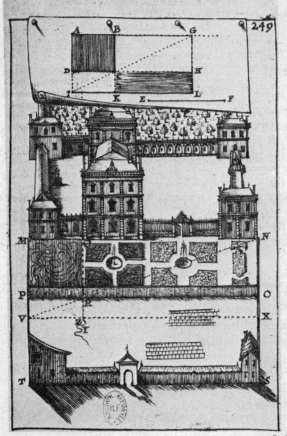

PLATE V. From Manesson Mallet's *Géométrie Practique* (1702).

PROPOSITION 44. *To a given straight line to apply, in a given rectilineal angle, a parallelogram equal to a given triangle.*

In modern terms: given one side and one angle construct a parallelogram equivalent to a given triangle.

PROPOSITION 45. *To construct, in a given rectilineal angle, a parallelogram equal to a given rectilineal figure.*

PROPOSITION 46. *On a given straight line to describe a square.*

The last two propositions, numbers 47 and 48, are Pythagoras's theorem and its converse. They will be discussed fully in Chapter 4, Vol. 2.

BOOK II. This book, which is very short and is entirely concerned with the equivalence of rectangles, is an important part of what has been called the 'geometrical algebra' of the Greeks.

We shall discuss it later in Chapter 3, Vol. 2.

BOOK III. This book is concerned exclusively with the circle. It contains eleven definitions and thirty-seven propositions.

DEFINITION 1. **Equal circles** are those the diameter of which are equal, or the radii of which are equal.*

DEFINITION 2. A straight line is said to **touch a circle** which, meeting the circle and being produced, does not cut the circle.

To understand the significance of this definition we must remember that the circle is a plane area circumscribed by a line, and that Euclid used the words 'straight line' where we should say 'segment of a straight line'.

---

* The word radius is used in the sense 'radius of a circle' by Cicero, and then by Pierre de la Ramée (Ramus) in his *Scholae Mathematicae* of 1569. Before the eighteenth century it was usual to use 'semi-diameter' or 'demi-diameter' instead of 'radius'. Some eighteenth-century writers also adopt this usage. The word *radian* to indicate the angle subtended at the centre by an arc of length equal to the radius of the circle, dates from the twentieth century.

DEFINITION 3. **Circles** are said to touch one another which, meeting one another, do not cut one another.

DEFINITION 6. A **segment of a circle** is the figure contained by a straight line and a circumference of a circle.*

DEFINITION 7. An **angle of a segment** is that contained by a straight line and a circumference of a circle.

It is thus an angle contained by a straight line and a curve.

DEFINITION 8. An **angle in a segment** is the angle which, when a point is taken on the circumference of the segment and straight lines are joined from it to the extremities of the straight line which is the **base of the segment**, is contained by the straight lines so joined.

DEFINITION 9. And, when the straight lines containing the angle cut off a circumference, the angle is said to **stand upon** that circumference.

DEFINITION 10. A **sector of a circle** is the figure which, when an angle is constructed at the centre of the circle, is contained by the straight lines containing the angle and the circumference cut off by them.

DEFINITION 11. **Similar segments of circles** are those which admit equal angles, or in which the angles are equal to one another.

PROPOSITION 1. *To find the centre of a given circle.*

The circle is presumed to be already drawn. We draw a chord, and then construct its perpendicular bisector. This is a diameter of the circle. We take its mid-point and prove by *reductio ad absurdum* that this is the required centre.

PORISM.† From this is manifest that, if in a circle a straight line cuts a straight line into two equal parts and at right angles, the centre of the circle is on the cutting straight line.

---

* The word 'periphery' ($\pi\epsilon\rho\iota\phi\epsilon\rho\epsilon\iota\alpha$), rendered in Latin as 'circumference', is used by Greek geometers where we should use the word 'arc'. It can however refer to a complete circumference.

Arab geometers use a word which can be literally translated as bow (Fr: arc). Campanus, whose *Euclid* is at least in part a translation from the Arabic, writes: *Portio circumferentiae, arcus nuncupatur.*

The word 'chord' does not exist in Greek: the term used is 'straight line in the circle'. The word 'chord', like the word 'arc', comes to us from the Arabs, and it seems to have been introduced into the West in the twelfth century AD by Plato of Tivoli.

† The modern term is 'corollary'.

PROPOSITION 2. *If on the circumference of a circle two points be taken at random, the straight line joining the points will fall within the circle.*

Proof by *reductio ad absurdum*

PROPOSITION 3. *If in a circle a straight line through the centre bisects a straight line not through the centre, it also cuts it at right angles; and if it cuts it at right angles, it also bisects it.*

PROPOSITION 4. *If in a circle two straight lines cut one another which are not through the centre, they do not bisect one another.*

PROPOSITION 5. *If two circles cut one another, they will not have the same centre.*

PROPOSITION 6. *If two circles touch one another, they will not have the same centre.*

PROPOSITION 7. *If on the diameter of a circle a point be taken which is not the centre of the circle, and from the point straight lines fall upon the circle, that will be greatest upon which the centre is, the remainder of the same diameter will be least, and of the rest the nearer to the straight line through the centre is always greater than the more remote, and only two equal straight lines will fall from the point on the circle, one on each side of the least straight line.*

PROPOSITION 8. An analogous proposition for a point outside the circle. Euclid takes care to distinguish between the concave part of the circumference (κοίλης περιφερείας) and the convex part (κυρτῆς περιφερείας).

PROPOSITION 9. *If a point be taken within a circle, and more than two equal straight lines fall from the point on the circle, the point taken is the centre of the circle.*

PROPOSITION 10. *A circle does not cut a circle at more points than two.*

PROPOSITION 11. *If two circles touch one another internally, and their centres be taken, the straight line joining their centres, if it be also produced, will fall on the point of contact of the circles.*

PROPOSITION 12, which is analogous to the preceding proposition, refers to external contact. It appears to be an interpolation.

PROPOSITION 13. *A circle does not touch a circle at more points than one, whether it touch it internally or externally.*

PROPOSITION 14. *In a circle equal straight lines are equally distant from the centre, and those which are equally distant from the centre are equal to one another.*

PROPOSITION 15. *Of straight lines in a circle the diameter is greatest, and of the rest the nearer to the centre is always greater than the more remote.*

PROPOSITION 16. *The straight line drawn at right angles to the diameter of a circle at its extremity will fall outside the circle, and into the space between the straight line and the circumference another straight line cannot be interposed; further the angle of the semicircle is greater, and the remaining angle less, than any acute rectilineal angle.*

The first part of the proposition establishes the existence of a tangent to the circle at any point on it. The second part proves that this tangent is unique. The third part is one of the few passages where Euclid considers angles contained by curves and straight lines. It gave rise to centuries of debate about the 'angle of contact' (the angle Euclid here calls the 'remaining angle', and which Proclus calls 'horn-shaped').

PROPOSITION 17. *From a given point to draw a straight line touching a given circle.*

To construct a tangent from the point *A* to the circle *BCD* Euclid proceeds as follows (Fig. 4.4): he draws the line *AE*,

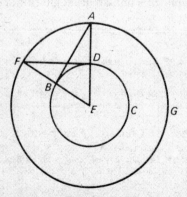

Fig. 4.4

which cuts the circle at *D*. He then constructs *DF* perpendicular to *AE*, and takes as *F* the point of intersection of the line *DF* with the circle centre *E* passing through *A*. *EF* cuts the given circle at *B*. *AB* is the required tangent.

In the *Data*, Proposition 91, Euclid obtains *B* in a different way: as the intersection of the given circle and the circle with diameter *AE*.

PROPOSITION 18. *If a straight line touch a circle, and a straight line be joined from the centre to the point of contact, the straight line so joined will be perpendicular to the tangent.*

PROPOSITION 19 is a converse of Proposition 18.

PROPOSITION 20. *In a circle the angle at the centre is double of the angle at the circumference, when the angles have the same circumference as base.*

PROPOSITION 21. *In a circle the angles in the same segment are equal to one another.*

PROPOSITION 22. *The opposite angles of quadrilaterals in circles are equal to two right angles.*

PROPOSITION 23. *On the same straight line there cannot be constructed two similar and unequal segments of circles on the same side.*

If not we should have a triangle in which an exterior angle was equal to a non-adjacent interior one.

PROPOSITION 24. *Similar segments of circles on equal straight lines are equal to one another.*

PROPOSITION 25. *Given a segment of a circle, to describe the complete circle of which it is a segment.*

PROPOSITION 26. *In equal circles equal angles stand on equal circumferences, whether they stand at the centres or at the circumferences.*

PROPOSITION 27. Converse of Proposition 26.

PROPOSITION 28. *In equal circles equal straight lines cut off equal circumferences, the greater equal to the greater and the less to the less.*

PROPOSITION 29. Converse of Proposition 28.

PROPOSITION 30. *To bisect a given circumference.*

PROPOSITION 31. *In a circle the angle in the semicircle is right, that in a greater segment less than a right angle, and that in a less segment greater than a right angle; and further the angle of the greater segment is greater than a right angle, and the angle of the less segment less than a right angle.*

The last two angles are contained by straight lines and curves.

PROPOSITION 32. *If a straight line touch a circle, and from the point of contact there be drawn across, in the circle, a straight line cutting the circle, the angles which it makes with the tangent will be equal to the angles in the alternate segments of the circle.*

PROPOSITION 33. *On a given straight line to describe a segment of a circle admitting an angle equal to a given rectilineal angle.*

This is the 'subtending segment' or 'subtending arc'.

PROPOSITION 34. *From a given circle to cut off a segment admitting an angle equal to a given rectilineal angle.*

PROPOSITION 35. *If in a circle two straight lines cut one another, the rectangle contained by the segments of the one is equal to the rectangle contained by the segments of the other.*

Let the chords be *AC* and *BD*, and let their point of intersection be *E* (see Fig. 4.5).

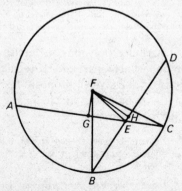

Fig. 4.5

Let the mid-points of the chords be G and H.

Now $AE . EC + GE^2 = GC^2$ (Book II, Proposition 5, see Vol. 2, p. 79)

$$\therefore \quad AE . EC + GE^2 + GF^2 = GC^2 + GF^2$$

and $AE . EC + FE^2 = FC^2$. Similarly,

$$BE . ED + FE^2 = BF^2.$$

But $BF = FC$,

$$\therefore \quad AE . EC = BE . ED.$$

PROPOSITION 36. *If a point be taken outside a circle and from it there fall on the circle two straight lines, and if one of them cut the circle and the other touch it, the rectangle contained by the whole of the straight line which cuts the circle and the straight line intercepted on it outside between the point and the convex circumference will be equal to the square on the tangent.*

This proposition is proved in the same way as the preceding one.

PROPOSITION 37. *If a point be taken outside a circle and from the point there fall on the circle two straight lines, if one of them cut the circle, and the other fall on it, and if further the rectangle contained by the whole of the straight line which cuts the circle and the straight line intercepted on it outside between the point and the convex circumference be equal to the square on the straight line which falls on the circle, the straight line which falls on it will touch the circle.*

BOOK IV. Inscribing polygons in circles and circumscribing them about them, particularly regular polygons.

Seven definitions and sixteen propositions, which are all problems; of the definitions, which fix the terminology, the last may perhaps seem strange:

DEFINITION 7. A straight line is said to be **fitted into a circle** [ἐναρμόζεσθαι, accomodari seu coaptari] when its extremities are on the circumference of the circle.

PROPOSITION 1. *Into a given circle to fit a straight line equal to a given straight line which is not greater than the diameter of the circle.*

PROPOSITION 2. *In a given circle to inscribe a triangle equiangular with a given triangle.*

PROPOSITION 3. Same data. The triangle is required to circumscribe the circle.

PROPOSITION 4. *In a given triangle to inscribe a circle.*

PROPOSITION 5. *About a given triangle to circumscribe a circle.*

PROPOSITIONS 6 AND 7. To inscribe a square in a given circle. To circumscribe a square about a given circle.

PROPOSITIONS 8 AND 9. To inscribe a circle in a given square. To circumscribe a circle about a given square.

PROPOSITION 10. *To construct an isosceles triangle having each of the angles at the base double of the remaining one.*

The proof is notable in that it avoids any appeal to similarity (Fig. 4.6):

Let a straight line AB be divided by the point C, such that

$$AC^2 = BC \cdot BA \text{ (Book II, Proposition 11, see Vol. 2, p. 84).}$$

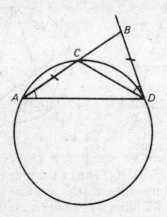

Fig. 4.6

The isosceles triangle ABC, which has equal sides AB and AD and base $BD = AC$, fulfils the requirements of the proposition.

Since $BC \cdot BA = AC^2 = BD^2$, BD is a tangent to the circle which circumscribes the triangle ACD. Therefore angle $C\widehat{D}B$ is equal to angle $C\widehat{A}D$.

The triangles ABD and CDB thus have two angles in common. Therefore they are equiangular.

Angle $B\widehat{C}D$ is thus equal to angle $C\widehat{B}D$, the triangle DBC is isosceles and $CD = BD$.

Now $BD = AC$, so $CD = AC$ and the triangle ACD is isosceles. The angle $C\widehat{A}D$ is equal to the angle $A\widehat{D}C$ and is thus one half of the size of the angles at the base of the triangle ABD.

PROPOSITION 11. *In a given circle to inscribe an equilateral and equiangular pentagon.*

We construct an isosceles triangle FGH whose base angles are equal to twice the angle at the vertex (Fig. 4.7).

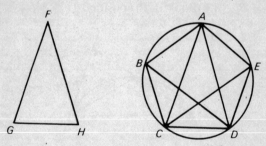

Fig. 4.7

In the given circle we inscribe a triangle ACD equiangular to the triangle FGH. If DB and CE are the bisectors of the base angles of the triangle ACD, it can be shown that the pentagon $ABCDE$ fulfils the required conditions.*

* See Vol. 2, p. 127 for Ptolemy's more elegant solution to this problem.

PROPOSITIONS 12, 13, 14. To circumscribe a regular pentagon about a circle. To inscribe a circle in a pentagon. To circumscribe a circle about a pentagon.

PROPOSITION 15. *In a given circle to inscribe an equilateral and equiangular hexagon.*

PROPOSITION 16. In a given circle to inscribe an equiangular and equilateral polygon with fifteen sides.

The side of a pentagon subtends $\frac{1}{5}$ (i.e. $\frac{3}{15}$) of the circumference of the circumscribing circle, while that of an equilateral triangle subtends $\frac{5}{15}$. The difference between these arcs is $\frac{2}{15}$, so we take half of it.

BOOK V. The first four books were elementary, but the fifth one is exceedingly advanced.

It deals with the theory of proportion, generalized to include cases where the magnitudes being compared are incommensurable with one another.

This book has been the subject of critical analysis for a period of two thousand years and in that time has given rise to an immensely rich literature, both among the Arabs and in the West. In the seventeenth century men like Galileo and Torricelli found it unsatisfactory, and only Barrow defended it. Not until the second half of the nineteenth century was it fully understood and accepted as the basis for further work.

Among the more important of the definitions at the beginning of the book we find [3]:

3. A **ratio** is a sort of relation in respect of size between two magnitudes of the same kind.

4. Magnitudes are said to **have a ratio** to one another which are capable, when multiplied, of exceeding one another.

5. Magnitudes are said to **be in the same ratio**, the first to the second and the third to the fourth, when, if any equimultiples whatever be taken of the first and third, and any equimultiples whatever of the second and fourth, the former equimultiples alike exceed, are alike equal to, or alike fall short of, the latter equimultiples respectively taken in corresponding order.

6. Let magnitudes which have the same ratio be called **proportional**.

7. When, of the equimultiples, the multiple of the first magnitude exceeds the multiple of the second, but the multiple of the third does not exceed the multiple of the fourth, then the first is said to **have a greater ratio** to the second than the third has to the fourth.

The fourth definition says that we may not speak of the ratio of two quantities of some particular type, two lengths for example, unless we prove or postulate the following property: If X and Y are two given arbitrary quantities of this type, then given an integer n it is always possible to find an integer m such that $mX \geqslant nY$. We shall find a similar postulate in the work of Archimedes.

The fifth definition would be expressed in modern terms as follows: If A and B are two quantities of the same type, satisfying Definition 4, and if C and D are two other quantities which also satisfy this definition, then the ratio between the quantities A and B is said to be equal to the ratio between the quantities C and D (i.e. $A:B = C:D$) if, for arbitrary integers m and n the inequality $mA > nB$ implies $mC > nD$, and the inequality $mA < nB$ implies $mC < nD$, and $mA = nB$ implies $mC = nD$.

DEFINITION 7. Unequal ratios: If there exists a pair of integers m and n such that $mA > nB$ and $mC \leqslant nD$, then the ratio $A:B$ is said to be greater than the ratio $C:D$.

Working from these three definitions Euclid proceeds, with almost flawless logic, to deduce twenty-five propositions which deal with the most important properties of ratios.*

BOOK VI. The purpose of this book is to apply to plane geometry the theoretical results obtained in Book V.

The first of the four definitions reads [4]:

1. **Similar rectilineal figures** are such as have their angles severally equal and the sides about the equal angles proportional.

---

* On the terminology for ratios and proportions see above pp. 44–47.

The book contains thirty-three propositions. In Vol. 2, Chapter 3 (pp. 85–92) we shall deal with those which provide geometrical methods of solving quadratic equations.

PROPOSITION 1. *Triangles and parallelograms which are under the same height are to one another as their bases.*

Let *ABC, ACD* be triangles and *EC, CF* parallelograms under the same height [Fig. 4.8];

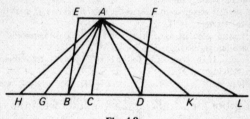

Fig. 4.8

I say that, as the base *BC* is to the base *CD*, so is the triangle *ABC* to the triangle *ACD*, and the parallelogram *EC* to the parallelogram *CF*.

For let *BD* be produced in both directions to the points *H, L* and let [any number of straight lines] *BG, GH* be made equal to the base *BC*, and any number of straight lines *DK, KL* equal to the base *CD*; let *AG, AH, AK, AL* be joined.

Then, since *CB, BG, GH* are equal to one another, the triangles *ABC, AGB, AHG* are also equal to one another.   [I. 38]

Therefore, whatever the multiple the base *HC* is of the base *BC*, that multiple also is the triangle *AHC* of the triangle *ABC*.

For the same reason, whatever multiple the base *LC* is of the base *CD*, that multiple also is the triangle *ALC* of the triangle *ACD*; and, if the base *HC* is equal to the base *CL*, the triangle *AHC* is also equal to the triangle *ACL*,   [I. 38]

if the base *HC* is in excess of the base *CL*, the triangle *AHC* is also in excess of the triangle *ACL*, and, if less, less.

Thus, there being four magnitudes, two bases *BC, CD* and two triangles *ABC, ACD*, equimultiples have been taken of the base *BC* and the triangle *ABC*, namely the base *HC* and the triangle *AHC*, and of the base *CD* and the triangle *ADC* other, chance, equimultiples, namely the base *LC* and the triangle *ALC*;

and it has been proved that,
if the base HC is in excess of the base CL, the triangle AHC is also in excess of the triangle ALC;
if equal, equal; and, if less, less.

Therefore, as the base BC is to the base CD, so is the triangle ABC to the triangle ACD.   [v. Def. 5]

Next, since the parallelogram EC is double of the triangle ABC,   [I. 41]
and the parallelogram FC is double of the triangle ACD,
while parts have the same ratio as the same multiples of them,   [v. 15]
therefore, as the triangle ABC is to the triangle ACD, so is the parallelogram EC to the parallelogram FC.

Since, then, it was proved that, as the base BC is to CD, so is the triangle ABC to the triangle ACD,
and, as the triangle ABC is to the triangle ACD, so is the parallelogram EC to the parallelogram CF,
therefore also, as the base BC is to the base CD, so is the parallelogram EC to the parallelogram FC.   [v. 11]

Therefore etc.   [Q.E.D.] [5]

The proof for parallelograms follows from the fact that each parallelogram is made up of two of the corresponding triangles.

PROPOSITION 2. *If a straight line be drawn parallel to one of the sides of a triangle, it will cut the triangle proportionally; and, if the sides of the triangle be cut proportionally, the line joining the points of section will be parallel to the remaining side of the triangle.*

Proof of the first part (see Fig. 4.9):

DE is parallel to BC

∴ the triangles BDE and CDE are equal

∴ $BDE : ADE = CDE : ADE.$

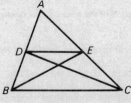

Fig. 4.9

But $BDE:ADE = BD:AD$, by the previous proposition. Similarly, $CDE:ADE = CE:EA$

$$\therefore \quad BD:AD = CE:EA.$$

The second part is proved by taking the same equations in the reverse order.

PROPOSITION 3. The internal bisector of an angle of a triangle divides the opposite side into segments proportional to the adjacent sides.

Euclid's proof is still regarded as the classic one.

PROPOSITION 4. *In equiangular triangles the sides about the equal angles are proportional, and there are corresponding sides which subtend the equal angles.*

The equiangular triangles ABC and DCE are placed as shown in Fig. 4.10. The pairs of corresponding sides are then parallel, and the proposition reduces to Proposition 2.

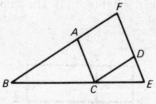

Fig. 4.10

PROPOSITION 5. Two triangles whose corresponding sides are proportional to one another are equiangular.

PROPOSITION 6. The same is true of two triangles in which two pairs of corresponding sides are proportional to one another and the angles enclosed between them are equal.

PROPOSITION 7. *If two triangles have one angle equal to one angle, the sides about the other angles proportional, and the remaining angles either both less or both not less than a right angle, the triangles will be equiangular and will have those angles equal, the sides about which are proportional.*

PROPOSITION 8. *If in a right-angled triangle a perpendicular be drawn from the right angle to the base, the triangles adjoining the perpendicular are similar both to the whole and to one another.*

PORISM. From this it is clear that, if in a right-angled triangle a perpendicular be drawn from the right angle to the base, the straight line so drawn is a mean proportional between the segments of the base.

PROPOSITIONS 9, 10, 11, 12. To find a part of a segment ($\frac{1}{3}$ in the example given), to divide a straight line into parts proportional to two given segments, to find a third and a fourth proportional.

PROPOSITION 13. To find a mean proportional.

PROPOSITIONS 14, 15, 16, 17. If two parallelograms are equal and equiangular the sides enclosing the equal angles are proportional to one another.

Converse (14) Let (*AB*) denote the area of the parallelogram with diagonal *AB* (Fig. 4.11). The parallelograms *AB* and *BC* are equal.

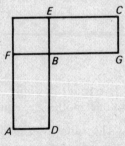

Fig. 4.11

Therefore

$$(AB):(FE) = (BC):(FE)$$

But

$$(AB):(FE) = BD:BE$$

and

$$(BC):(FE) = BG:FB,$$

therefore

$$BD:BE = BG:FB.$$

Q.E.D.

The converse is proved by taking the above equations in the reverse order.

(15) proves the analogous proposition for triangles

(16) is the analogous proposition for rectangles, equivalent to:

If $a:b = c:d$ then $ad = bc$, and vice versa.

(17) shows similarly that:

If $a:b = b:c$ then $b^2 = ac$, and vice versa.

The products represent rectangles with sides of the appropriate lengths.

PROPOSITIONS 18, 19, 20. Similar polygons are in a ratio double that of the corresponding sides. (In modern terms this would read: the ratio of the areas of two similar polygons is the square of the ratio of their corresponding sides.)

PROPOSITION 21. *Figures which are similar to the same rectilineal figure are also similar to one another.*

PROPOSITION 22. *If four straight lines be proportional, the rectilineal figures similar and similarly described upon them will also be proportional;*

and the converse

PROPOSITION 23. *Equiangular parallelograms have to one another the ratio compounded of the ratio of their sides.*

The proof is like that of Proposition 14.

The following propositions consist of geometrical methods of solving quadratic equations. We shall consider them later, when we come to discuss such methods (see Vol. 2, pp. 85–92).

PROPOSITION 31. *In right-angled triangles the figure on the side subtending the right-angle is equal to the similar and similarly described figures on the sides containing the right-angle.*

The proof uses Propositions 8 and 19 of Book VI.

PROPOSITION 33. *In equal circles angles have the same ratios as the circumferences on which they stand, whether they stand at the centres or at the circumferences.*

The proof is like that of Proposition 1 of Book VI.

THE THREE ARITHMETICAL BOOKS:
Books VII, VIII and IX

BOOK VII. We have already mentioned (p. 37) some of the twenty-two definitions to be found at the beginning of Book VII.

The book contains thirty-nine propositions. The first nineteen deal with ratios and proportions and are thus analogous to those of Book V, but all the ratios considered in Book VII are rational, and can therefore be expressed as ratios of integers. Among these propositions there is a discussion of the highest common factor.

Propositions 20 to 26 concern numbers that are prime to one another, and Propositions 29 to 32 numbers that are absolutely prime.

There follows a discussion of the lowest common multiple.

The most important of the thirty-nine propositions in this book are the following:

PROPOSITION 1. *Two unequal numbers being set out, and the less being continually subtracted from the greater, if the number which is left never measures the one before it until a unit is left, the original numbers will be prime to one another.*

The operation thus defined is now known as 'Euclid's Algorithm'. It plays a fundamental part in the theory of numbers.

The wording is a little obscure. It can be paraphrased as: Given two numbers, A and B, of which A is the larger. We subtract B from A, and call the remainder C. We then subtract the smaller of the two numbers B and C from the larger one, and we continue this process until we obtain either a one or a zero. A number which when subtracted several times from another gives a zero remainder is said to 'measure it exactly'.

Euclid's Algorithm is used again in the second proposition:

PROPOSITION 2. *Given two numbers not prime to one another, to find their greatest common measure.*

Propositions on numbers which are prime to one another:

PROPOSITION 20. *The least numbers of those which have the same ratio with them measure those which have the same ratio the same number of times, the greater the greater and the less the less.*

That is, if a and b are the smallest integers which are in a given ratio then if the integers A and B are such that $A:B = a:b$ there exists an integer such that $A = ma$ and $B = mb$.

PROPOSITION 21. *Numbers prime to one another are the least of those which have the same ratios with them.*

PROPOSITION 22. Converse of Proposition 21.

PROPOSITION 23. *If two numbers be prime to one another, the number which measures the one of them will be prime to the remaining number.*

PROPOSITION 24. *If two numbers be prime to any number, their product also will be prime to the same.*

PROPOSITION 25. If two numbers are prime to one another, the square of one of them is prime to the other.

PROPOSITION 26. *If two numbers be prime to two numbers, both to each, their products will also be prime to one another.*

PROPOSITION 27. *If two numbers be prime to one another, and each by multiplying itself make a certain number, the products will be prime to one another; and, if the original numbers by multiplying the products make certain numbers, the latter will also be prime to one another [and this is always the case with the extremes].*

PROPOSITION 28. Two numbers which are prime to one another are prime to their sum, and vice versa.

Propositions on absolute prime numbers:

PROPOSITION 29. *Any prime number is prime to any number which it does not measure.*

PROPOSITION 30. *If two numbers by multiplying one another make some number, and any prime number measure the product, it will also measure one of the original numbers.*

PROPOSITION 31. *Any composite number is measured by some prime number.*

PROPOSITION 32. *Any number either is prime or is measured by some prime number.*

BOOK VIII. This book is concerned with establishing the conditions for the existence of rational nth roots (square roots, cube roots, etc.) of integers and fractions (we are here using the modern terminology).

There are several logical weaknesses in Book VII, but in Book VIII, which contains twenty-seven propositions and no definitions, only Propositions 22 and 23 are proved incorrectly. Correct proofs of these were supplied by Clavius at the end of the sixteenth century.

Euclid frequently refers to numbers being 'successively proportional' or in 'continued proportion'. In modern terms we would describe such numbers as a 'geometrical progression'.

Propositions 1, 2 and 3 show how to construct progressions with a given common ratio, in the lowest possible terms, that is using the smallest possible integers. The common ratio is reduced to its lowest terms a/b, and the numbers

$$a^n, a^{n-1}b, a^{n-2}b^2, \ldots, a^2b^{n-2}, ab^{n-1}, b^n,$$

then give the required progression.

The following problem, taken from the *Divisions*, a work ascribed to Euclid, shows how this technique was applied in the theory of music:

9. *Six sesquioctaval intervals are larger than a double interval.*

Musical intervals correspond to ratios of numbers, the sesquioctaval interval (see p. 45) is given by the ratio $9:8$. The double interval is given by $2:1$. The complete interval, the sum of the intervals, corresponds to the compound ratio, which we should now consider not as a sum but as a product of ratios.

Proof from the *Divisions*:

Let A be a number and let B be the sesquioctave of A, C the sesquioctave of B, D the sesquioctave of C, E the sesquioctave of D, F the sesquioctave of E and G the sesquioctave of F. I claim that G is greater than double A.

Since we are looking for seven numbers which are successively sesquioctaves of one another, let the series of numbers A, B, C, D, E, F and G start with $A = 262\,144[8^6]$.* Then

$$B = 294\,912[8^5 \times 9], \qquad C = 331\,776[8^4 \times 9^2],$$
$$D = 373\,248[8^3 \times 9^3], \qquad E = 419\,904[8^2 \times 9^4],$$
$$F = 472\,392[8 \times 9^5], \qquad G = 531\,441[9^6]$$

and G is more than double A.

The sixth proposition of Book VIII, is of fundamental importance.

---

* We use [ ] to indicate the factorized form of the numbers.

PROPOSITION 6. *If there be as many numbers as we please in continued proportion, and the first do not measure the second, neither will any other measure any other.*

Let the numbers in question be A, B, C, D and E, so we have

$$A:B = B:C = C:D = D:E.$$

A does not measure B, which means that the ratio $A:B$, reduced to its lowest terms, is not an integer. Therefore none of the numbers measures the one that follows it. Also, A does not measure C for if $B:A = m:n$, where m and n are prime to one another, then m^2, mn, n^2 is a series with the required ratio and $A:C = m^2:n^2$, where m^2 and n^2 are prime to one another. But, if A measures C, then $A:C = 1:p$ and m^2 and n^2 are not the smallest integers which define the ratio $A:C$; which contradicts our original assumption. This argument clearly applies to any two terms of the progression.

Proposition 6 and its consequences thus achieve the purpose Theaetetus described to Socrates in the dialogue we quoted on p. 58.

BOOK IX. This book is much less homogeneous than the eighth. It contains thirty-six propositions. Some of the ones concerned with odd and even numbers, which we quoted on pages 37 to 41, seem to be very archaic. Others are both elegant and profoundly important. For example:

PROPOSITION 20. *Prime numbers are more than any assigned multitude of prime numbers.*

The proof is accepted as classical.

PROPOSITION 36. *If as many numbers as we please beginning from an unit be set out continuously in double proportion, until the sum of all becomes prime, and if this sum multiplied into the last make some number, the product will be perfect.*

Perfect numbers are defined at the beginning of Book VII:

DEFINITION 22. **A perfect number** is that which is equal to its own parts.

We should now express this by saying that such a number is equal to the sum of its divisors, including unity but excluding the number itself.

For example 6 is a perfect number: its divisors are 1, 2 and 3, and $1 + 2 + 3 = 6$.

Proposition 36 asserts that if $(2^n - 1)$ is prime then $2^{n-1}(2^n - 1)$ is perfect.

All such perfect numbers are even. In a work published posthumously in 1849 Euler proved the converse of this proposition: that all even perfect numbers are of the form given by Euclid.

There is no known odd perfect number. Very few even perfect numbers are known.

BOOK X. Volume 2, Chapter 3 (pp. 95 to 104) contains a summary of the contents of this book, which is the longest of the books of the *Elements*.

Here we shall merely quote the first two propositions:

PROPOSITION 1. *Two unequal magnitudes being set out, if from the greater there be subtracted a magnitude greater than its half, and from that which is left a magnitude greater than its half, and if this process be repeated continually, there will be left some magnitude which will be less than the lesser magnitude set out.*

PROPOSITION 2. *If, when the less of two unequal magnitudes is continually subtracted in turn from the greater, that which is left never measures the one before it, the magnitudes will be incommensurable.*

This amounts to a generalized restatement of Euclid's Algorithm, which was defined at the beginning of Book VII.

BOOK XI. This book is concerned with solid geometry.

The book contains thirty-nine propositions and starts with twenty-eight definitions. The last set, which refer to regular polyhedra, have already been discussed on pp. 48–49 above. The most important of the remaining definitions are:

1. A **solid** is that which has length, breadth, and depth.

2. An extremity of a solid is a surface.

3. A **straight line** is **at right angles to a plane**, when it makes right angles with all the straight lines which meet it and are in the plane.

4. A **plane** is **at right angles to a plane** when the straight lines drawn, in one of the planes, at right angles to the common section of the planes are at right angles to the remaining plane.

6. The **inclination of a plane to a plane** is the acute angle contained by the straight lines drawn at right angles to the common section at the same point, one in each of the planes.

8. **Parallel planes** are those which do not meet.

9. **Similar solid figures** are those contained by similar planes equal in multitude.

10. **Equal and similar solid figures** are those contained by similar planes equal in multitude and in magnitude.

11. A **solid angle** is the inclination constituted by more than two lines which meet one another and are not in the same surface, towards all the lines.

Otherwise: A **solid angle** is that which is contained by more than two plane angles which are not in the same plane and are constructed to one point.

14. When, the diameter of a semicircle remaining fixed, the semicircle is carried round and restored again to the same position from which it began to be moved, the figure so comprehended is a **sphere**.

18. When, one side of those about the right angle in a right-angled triangle remaining fixed, the triangle is carried round and restored again to the same position from which it began to be moved, the figure so comprehended is a **cone**.

And, if the straight line which remains fixed be equal to the remaining side about the right angle which is carried round, the cone will be **right-angled**; if less, **obtuse-angled**; and if greater, **acute-angled**.

Euclid's is not the only ancient definition of the sphere. In the fifth century BC, Parmenides wrote of the Whole:

Since it is perfect within an outer limit it is like a solid sphere curving outwards, equally distant from its centre at every point.

A definition similar to that given by Parmenides is to be found in the *Sphaerica* of Theodosius and in the *Definitions* of Heron.*

---

* Compare Cicero, *Universitate de mundo*: 'Ergo globulus est fabricatus, quod σφαιρῶειδες Graeci vocant, cujus omnis extremitas paribus a medio radiis attingitur.' For the use of the word 'radius' see p. 80, footnote.

The classification of cones of revolution given in Definition 18 helps us to understand the names Archimedes uses for the various conic sections.

Propositions:

1. *A part of a straight line cannot be in the plane of reference and a part in a plane more elevated.*

2. *If two straight lines cut one another, they are in one plane, and every triangle is in one plane.*

3. *If two planes cut one another, their common section is a straight line.*

The proofs given for these three propositions are inadequate: the axioms and postulates at the beginning of Book I do not enable us to deduce the propositions. Euclid has, in fact, introduced new postulates without stating them explicitly.

Propositions 4 and 5 concern the perpendicular to a plane.

4. *If a straight line be set up at right angles to two straight lines which cut one another, at their common point of section, it will also be at right angles to the plane through them.*

The accepted modern proof of this proposition is analogous to that given by Euclid. The figure, however, is different. We show here (Fig. 4.12) the figure appropriate to the *Elements*.*

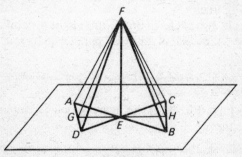

Fig. 4.12

---

* The modern figure is not found before 1830. Jean Ciermans (1640), Tacquet (1654) and Legendre (1794) use Pythagoras's Theorem.

5. If a straight line be set up at right angles to three straight lines which meet one another, at their common point of section, the three straight lines are in one plane.

The next group of propositions refer to parallelism:

6. Two lines perpendicular to the same plane are parallel to one another.

7. If two lines are parallel to one another any straight line joining a point of one to a point of the other lies in the plane of the two lines.
(The proof of this is inadequate. The proposition is really a postulate.)

8. Converse of 6.

9. Two lines which are both parallel to a third line are parallel to one another. (Proof by 6 and 8.)

10. The angles enclosed by intersecting pairs of parallel lines are equal. (The proof is accepted as classical.)

11. To drop a perpendicular to a plane from a point outside it or (12) to construct a perpendicular to a plane from a point on it.

13. Through a given point we can only draw one line perpendicular to a given plane.

14. Two planes perpendicular to the same line are parallel to one another.

15. The two angles in Proposition 10 lie in parallel planes.

16. A third plane cuts two parallel planes in parallel lines.

17. If two straight lines be cut by parallel planes, they will be cut in the same ratios.

We show (Fig. 4.13) the figure which accompanies this proposition.

18. If a straight line be at right angles to any plane, all the planes through it will also be at right angles to the same plane.

19. The intersection of two planes perpendicular to a third plane is also perpendicular to the third plane.

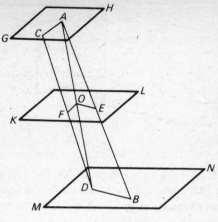

Fig. 4.13

There follow four propositions on polyhedral angles.

20. *If a solid angle be contained by three plane angles, any two, taken together in any manner, are greater than the remaining one.*
21. *Any solid angle is contained by plane angles less than four right angles.*

22, 23. To construct a triangular prism, given its three faces; and the conditions for this to be possible.

Propositions 24 to 38 are on parallelepipeds, and are mainly concerned with equivalence.

39. *If there be two prisms of equal height, and one have a parallelogram as base and the other a triangle, and if the parallelogram be double of the triangle, the prisms will be equal.*

Figure 4.14 shows such prisms. The two equal 'heights' are the distance from *BE* to the plane *ACDF* and the distance between the planes *HGK* and *LMN*.

The parallelogram *ADFC* has an area twice that of the triangle *HGK*.

This proposition is used in Book XII in Euclid's discussion of the volume of a pyramid. (See above p. 61.)

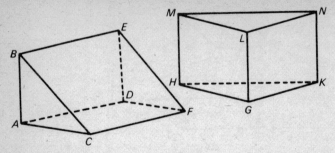

Fig. 4.14

BOOK XII. According to Archimedes, the results in this book were first obtained by Eudoxus. We discussed it in the previous chapter, on pp. 59 to 65. It contains eighteen propositions.

BOOK XIII. Book XIII deals with constructing the regular polyhedra we discussed above on pp. 48 to 54 (see Plate III). It contains eighteen propositions, some of them very long. Today they might provide material for interesting monographs in descriptive geometry.

Many editions of Euclid contain a further two books numbered XIV and XV, which continue the study of polyhedra. Book XV was written long after the other books, and part of it dates from the sixth century AD, being the work of an anonymous pupil of Isidore of Miletus, one of the two architects of the church of Santa Sophia* in Constantinople. Book XIV is the work of Hypsicles of Alexandria, who lived in the second century BC.

The Preface reads as follows:

'Basilides of Tyre, O Protarchus, when he came to Alexandria and met my father, spent the greater part of his sojourn with him on account of the bond between them due to their common interest in mathematics. And on one occasion, when looking into the tract written by Apollonius about the comparison of the dodecahedron and icosahedron inscribed in one and the

---

* Church of the Holy Wisdom. J.V.F.

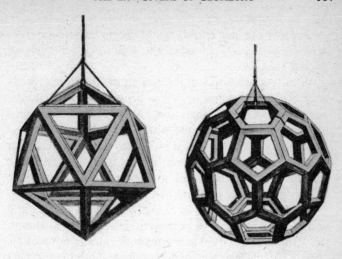

PLATE VI. (a) Regular, and (b) semi-regular polyhedra, drawn by Leonardo da Vinci for *De Divina Proportione* (1509) of Pacioli.

B.N.

same sphere, that is to say, on the question what ratio they bear to one another, they came to the conclusion that Apollonius' treatment of it in this book was not correct; accordingly, as I understood from my father, they proceeded to amend and rewrite it. But I myself afterwards came across another book published by Apollonius, containing a demonstration of the matter in question, and I was greatly attracted by his investigation of the problem. Now the book published by Apollonius is accessible to all; for it has a large circulation in a form which seems to have been the result of later careful elaboration.'

'For my part, I determined to dedicate to you what I deem to be necessary by way of commentary, partly because you will be able, by reason of your proficiency in all mathematics and particularly in geometry, to pass an expert judgment upon what I am about to write, and partly because, on account of your intimacy with my father and your friendly feeling towards myself, you will lend a kindly ear to my disquisition. But it is time to have done with the preamble and to begin my treatise itself.'

4.2. Archimedes

Montucla in his *History of Mathematics* (*Histoire des Mathématiques*', 1758) gives the following summary of Archimedes's life:

'While Astronomy was flourishing in Alexandria, Sicily gave birth to a geometer whose genius was to excite the admiration of future generations: Archimedes, whose name is familiar to all acquainted with History or with Science. A certain Heraclides wrote a biography of Archimedes, but this work, which would have been of great interest to us, has unfortunately been lost, and all we now know about Archimedes is contained in the brief outline which follows.

Archimedes was born in Syracuse in about 287 BC and according to Plutarch came from the same family as Hieron the King of that city. Since Archimedes owes none of his fame to having been born of noble lineage (for he would have been forgotten had he been an ordinary man) we shall not discuss the matter further, nor shall we insist on Cicero's view that Archimedes was "humilis homo". If it were indeed true that the Roman Orator, in one of those moments when, as often happened, he was carried away by the enthusiasm of his art, did speak disdainfully of Archimedes, the fact would not be important, for it would not influence any person of discernment. However, in various other passages, Cicero shows so much admiration for Archimedes that we can be sure that even in this passage his words were not intended to be disparaging. If he had regarded Archimedes as a commonplace man, would he have taken the trouble to search out his tomb when he visited the city of Syracuse? And having found it, would he have pointed it out to Archimedes's countrymen, reproaching them with their forgetfulness and indifference towards a man who had brought honour to his city?

'Archimedes studied all branches of Mathematics, but his genius is most apparent in his work in Geometry and in

Mechanics. He was so interested in these Sciences that he forgot to eat or drink, and his servants had to remind him, or sometimes almost force him, to satisfy these human needs. There are in our own time some examples of such absentmindedness, occasioned by intense concentration on some particular subject. Plutarch tells several other stories about Archimedes, but I shall omit them since I believe them to be apocryphal and more likely to excite ridicule than a proper respect for this great man. Although most of his work was motivated by practical considerations Archimedes always regarded practical work as merely a lowly servant of theory: he considered that all the ingenious machines he designed, to defend his city or for some other purpose, were only Geometrical exercises, not worth describing in writing. We must regret this attitude, since it deprived us of so many inventions, now vanished without trace, but on the other hand, it gives us an answer to the arguments of those we hear today inveighing against theoretical work and speaking of it almost as if it were idle curiosity. What better example is there to refute their assertions than that of Archimedes: a man who was at once a most ingenious inventor and a great theoretician? ...'

Montucla continues:

'... Hieron, King of Syracuse, set Archimedes a problem which led him to his discoveries in hydrostatics. The King had given a craftsman a certain amount of gold to make into a crown, but the craftsman was dishonest and had kept some of the gold for himself, replacing it with an equal weight of silver. The fraud was suspected, but the crown was a fine piece of workmanship and no-one wanted to damage it, so Archimedes was consulted about whether there was some way of finding out how much silver had been substituted for gold. He thought about the problem, and it is said the answer came to him when he was in his bath, whereupon he got out, delightedly shouting εύρηκα, εύρηκα,

I've found it, I've found it—a word he made famous. He is then alleged to have run naked through the streets of Syracuse repeating this word. Such stories are believed by common people because they like to believe great men are ridiculous, by way of revenge for being forced to acknowledge that they are superior: but judicious critics will not believe in either the miraculous deeds or the ridiculous behaviour ascribed to great men.

.   .   .   .   .

The Ancients attributed forty mechanical inventions to Archimedes; but only a few are hinted at in the writings that have come down to us. One such invention is the inclined screw, a strange device in which a body's own propensity to fall seems to be used to make it rise. The device is known as the Archimedean screw. Diodorus reports that Archimedes invented it when he was in Egypt, to help the Egyptians raise the water which gathered in depressions after the flooding of the Nile. According to Athenaeus sailors too believed Archimedes invented this device, which they used to bale out water from the holds of ships. It has been said that Archimedes invented the endless screw and was the first to use multiple pulleys, and he was, perhaps, the first to use a moving pulley, for no description of such an arrangement is to be found in Aristotle's Mechanics.

'When Hieron expressed his amazement at the marvellous things Archimedes had achieved by his mechanical inventions, the latter is said to have made the famous reply: "Give me where to stand and I will move the earth." According to principles known at the time Archimedes might indeed have imagined a machine such that, with its aid, a given force however small would be able to overcome any resistance however great. According to Pappus, such a machine was the fortieth of Archimedes's inventions* and it is said he demonstrated its power by using it to launch a huge ship all on his own. This anecdote must be fictional, as anyone will realize who is aware of the effect of friction on the working of machines. Moreover, it is a principle of mechanics that what one gains in force one loses in time or in velocity. A man who uses a machine to perform a task which would otherwise require the natural forces of a hundred men, will take one hundred times as long to perform the task; so Archimedes would have taken a considerable time to make such an enormous load move forward. Several poets have written about one of the most famous of Archimedes's inventions, an instrument known as the Archimedean sphere, which shows the movements of heavenly bodies.

'Cicero also speaks of it with admiration, and regards it as one of mankind's greatest achievements. This particular invention was also the one with which Archimedes himself was most pleased: for although he had not bothered to describe his other inventions he did leave a description of this one, under the title of *Sphaeropoeia*. The work has unfortunately been lost.

'Tertullian seems to believe that Archimedes was the first to construct a hydraulic organ, an achievement usually ascribed to Ctesibius. But should we attach great weight to the evidence of this Father of the Church, who is certainly venerable in other respects but does not have the same authority in matters such as these? The grammarian Atilius Fortunatianus speaks of an invention which we shall not attempt to describe in terms other than his own, which we confess we do not understand.† . . .

---

* Montucla used Commandino's translation of Pappus's *Collection*. At this point there is an error in the translation, which has given rise to the legend of Archimedes's *forty* mechanical inventions.

† This is the *Stomachion* or *Loculus Archimedius* to which we shall refer on p. 117. The description is correct.

'Finally, we must mention how Archimedes used his skill in mechanics to defend his country: for it was in this that he showed his genius to the full, and demonstrated the great power of mathematics. Hieron's successor had ill-advisedly quarrelled with the Romans, and they seized their opportunity to gain control of Sicily. In 212 BC, after several military successes, they laid seige to Syracuse. The citizens were terrified by what they had seen and heard of Roman military prowess and would have put up little resistance had not Archimedes raised their spirits and organized one of the most vigorous defensive actions in recorded history. A number of machines, each more effective than the last, soon put a stop to the projects of the Roman engineers; for all their courage, the soldiers could not withstand the threatening aspect of these machines and were so afraid that they drew back or refused to advance.

'Marcellus was forced to give up the idea of taking the city by storm, and abandoned the seige in favour of a blockade. Descriptions of Archimedes's machines may be found in the works of Polybius, Livy and Plutarch, or in Chevalier Folard's commentary on the first of these authors. Since these works are freely available I shall not discuss the matter further here.

'This is the natural point at which to discuss the famous story of the burning glasses, which, it is said, Archimedes used to set fire to the Roman fleet. The story is based on reports by Zonaras and Tzetzes: the former relies on the testimony of Dion and the latter on that of Diodorus, Dion and several other writers. However, if we come to examine it in detail the story is open to so many objections that we can hardly be surprised that, despite the many authorities quoted, scholars have been divided in their opinion of its truth.

.   .   .   .   .   .

'We have mentioned that the Syracusans' resistance was such that Marcellus discontinued the seige. He instituted a blockade instead, and waited for some occasion when he might take the city by surprise. The confidence of the Syracusans soon gave him his chance. One day, occupied with celebrations of a festival of Diana and thinking that the Romans' morale had been so broken by their losses that they would not mount an attack, the Syracusans left the walls unguarded. The Romans noticed this, and quickly launched the scaling operation for which they had long been prepared. They entered the city, which was soon taken and sacked. It is said that Archimedes was so absorbed in his studies that he did not notice when a Roman soldier entered his room. Marcellus greatly admired this extraordinary man and had given orders that he should be spared, but his orders were not carried out. Perhaps the unfortunate mathematician was so occupied with his thoughts that the soldier lost patience with him, or perhaps the soldier was tempted by the apparently valuable nature of the contents of a box he was carrying, but, for whatever reason, Archimedes was killed, and did not survive the capture of his native city. This happened in the year 542 from the foundation of Rome, 212 BC.

'Valerius Maximus says that Marcellus deeply regretted the death of Archimedes and since it was now too late to save him he showed his generosity towards his dependants, showering favours upon those who had escaped the soldiers' anger. He returned their belongings to them, and also Archimedes's body, so that he might be given a fitting burial. Archimedes had expressed the wish that his tomb should bear an inscription showing a sphere drawn inside a cylinder, in memory of his discovery of the relation between these two bodies. His wish was complied with, and it was from this inscription that Cicero, when he was a quaestor in Sicily, recognized Archimedes's monument among the brambles and thorn bushes that hid it from sight.'

The following are what modern critics consider to be Archimedes's works, listed in approximately chronological order:

1. ON PLANE EQUILIBRIUMS, BOOK I

Ἐπιπέδων ἰσορροπιῶν ἤ κέντρα βαρῶν ἐπιπέδων α΄

De planorum aequilibriis sive de centris gravitatis planorum I.

No preface. Seven hypotheses. Fifteen propositions on the centres of gravity of parallelograms and triangles.

2. QUADRATURE OF THE PARABOLA

Τετραγωνισμὸς παραβολῆς

Quadratura Parabolae.

The title is apocryphal. The word 'parabola' is never used by Archimedes to denote a particular second-degree curve.

A preface and twenty-four propositions.

Later we shall present a brief description of the methods used in this work.

3. ON PLANE EQUILIBRIUMS, BOOK II

No preface. Ten propositions on the centres of gravity of segments of a parabola.

4. ON THE SPHERE AND CYLINDER
BOOKS I AND II

Περὶ σφαίρας καὶ κυλίνδρου α΄β΄

De Sphaera et Cylindro.

A preface, six "axioms" or definitions. Five postulates. Forty-four propositions in Book I. We shall discuss this book briefly later.* In Book II there are a preface and nine propositions.

5. ON SPIRALS

Περὶ ἑλίκων

De lineis spiralibus.

A preface, twenty-eight propositions on the Archimedean Spiral ($\rho = a\omega$), the locus of a point which moves uniformly along a straight line which itself turns with uniform angular velocity about a fixed point. Discussion of tangents to the curve and of areas swept out by the radius vector.

6. ON CONOIDS AND SPHEROIDS

Περὶ κωνοειδέων καὶ σφαιροειδέων

De Conoidibus et Sphaeroidibus.

A preface, some definitions, thirty-two propositions. Discussion of the volumes swept out by ellipses and parabolas rotating about an axis of symmetry and by hyperbolae rotating about a transverse axis.

7. MEASUREMENT OF A CIRCLE

Κύκλου μέτρησις

Dimensio Circuli.

---

* Montucla says of this book: 'These discoveries about the relation between the sphere and the cylinder pleased Archimedes so much that he wished the figures to be carved on his tomb. As we shall see, this was in fact done.' He adds in a note that 'Archimedes is not the only person who wished for such an epitaph to pass his discoveries on to posterity'.

'Ludolph van Ceulen wished to have carved on his tomb the famous numbers he had derived as limits for the length of the circumference of a circle. The project was executed, and the geometrical monument can still be seen, or so I have read somewhere. Jacques Bernoulli was so pleased with what he had discovered about the logarithmic spiral that he is said to have wished this curve to be carved on his tomb, with the words *eadem mutata resurgo*, which allude to some of the interesting properties of this spiral, as we shall explain later.'

In Chapter 7, Vol. 2, p. 186, there is a translation of this short treatise.

8. THE SAND-RECKONER

Ψαμμίτης

Arenarius.

This is a fanciful exposition of a system for counting very large numbers. The theme is the calculation of the number of grains of sand which would fill the sphere of the World. Archimedes explains, among other things, the theory of Aristarchus of Samos, that the earth moved round the sun.

9. ON FLOATING BODIES. BOOKS I AND II

'Οχουμένων α'β'

De corporibus fluitantibus.

Book I. One hypothesis.—Nine propositions, mainly dealing with general matters, among them Archimedes' Principle.

Book II. Ten propositions on the equilibrium of a homogeneous segment of a paraboloid of revolution floating in a liquid.

10. THE METHOD

Περὶ τῶν μηχανικῶν θεωρημάτων πρὸς 'Ερατοσθένην ἔφοδος

De mechanicis propositionibus ad Eratosthenem methodus.

A preface, eleven statements of lemmas,* fourteen propositions.

In this very important treatise, which was assumed lost until a copy came to light in Istanbul in 1906, Archimedes explains to Eratosthenes some of the methods he employed.

---

* Lemma: a proposition that is assumed or demonstrated, preliminary to the demonstration of some other proposition.

To these we may add the *Stomachion* or *Loculus Archimedius*, an entertaining little geometrical treatise about cutting up a rectangular plate, the *Lemmas* or *Liber Assumptorum*, a collection of thirteen geometrical propositions which has come down to us only in an Arabic translation, *The Cattle Problem* (an interesting quadratic problem in indeterminate analysis leading to a Fermat equation whose solution involves exceedingly large numbers) and, finally, there is a treatise on semi-regular polyhedra. Some of these Archimedean bodies are to be found in the *De divina proportione* of Pacioli (see Plate VI, p. 107).

Father Tacquet, who turned the *Treatise on the Sphere and Cylinder* into a school textbook (at the cost of some distortion), wrote in 1654 that those who praised Archimedes were more numerous than those who read him, and those who admired him were more numerous than those who understood him.

Here we shall consider only the quadrature of the parabola and the formulae for the volume and area of the sphere. We shall begin with the preface to *The Method*.

'Archimedes to Eratosthenes greeting.

I sent you on a former occasion some of the theorems discovered by me, merely writing out the enunciations and inviting you to discover the proofs, which at the moment I did not give. The enunciations of the theorems which I sent were as follows.

1. If in a right prism with a parallelogrammic base a cylinder be inscribed which has its bases in the opposite parallelograms,* and its sides [i.e. four generators] on the remaining planes (faces) of the prism, and if through the centre of the circle which is the base of the cylinder and (through) one side of the square in the plane opposite to it a plane be drawn, the plane so drawn will cut off from the cylinder a segment which is bounded by two planes and the surface of the cylinder, one of the two planes being the plane which has been drawn and the other the plane in which the base of the cylinder is, and the surface being that which is between the said planes; and the segment cut off from the cylinder is one sixth part of the whole prism.

---

* The parallelograms are apparently *squares*.

2. If in a cube a cylinder be inscribed which has its bases in the opposite parallelograms* and touches with its surface the remaining four planes (faces), and if there also be inscribed in the same cube another cylinder which has its bases in other parallelograms and touches with its surface the remaining four planes (faces), then the figure bounded by the surfaces of the cylinders, which is within both cylinders, is two-thirds of the whole cube.

Now these theorems differ in character from those communicated before; for we compared the figures then in question, conoids and spheroids and segments of them, in respect of size, with figures of cones and cylinders: but none of those figures have yet been found to be equal to a solid figure bounded by planes; whereas each of the present figures bounded by two planes and surfaces of cylinders is found to be equal to one of the solid figures which are bounded by planes. The proofs then of these theorems I have written in this book and now send to you. Seeing moreover in you, as I say, an earnest student, a man of considerable eminence in philosophy, and an admirer [of mathematical inquiry], I thought fit to write out for you and explain in detail in the same book the peculiarity of a certain method, by which it will be possible for you to get a start to enable you to investigate some of the problems in mathematics by means of mechanics. This procedure is, I am persuaded, no less useful even for the proof of the theorems themselves; for certain things first became clear to me by a mechanical method, although they had to be demonstrated by geometry afterwards because their investigation by the said method did not furnish an actual demonstration. But it is of course easier, when we have previously acquired, by the method, some knowledge of the questions, to supply the proof than it is to find it without any previous knowledge. This is a reason why, in the case of the theorems the proof of which Eudoxus was the first to discover, namely that the cone is a third part of the cylinder, and the pyramid of the prism, having the same base and equal height, we should give no small share of the credit to Democritus who was the first to make the assertion with regard to the said figure† though he did not prove it. I am myself in the position of having first made the discovery of the theorem now to be published [by the method indicated], and I deem it necessary to expound the method partly because I have already spoken of it‡ and I do not want to be thought to have uttered vain words, but equally because I am persuaded that it will be of no little service to mathematics; for I apprehend that some, either of my contemporaries or of my successors, will, by means of the method when once established, be able to discover other theorems in addition, which have not yet occurred to me.

---

* i.e. squares.

† περὶ τοῦ εἰρημένου σχήματος, in the singular. Possibly Archimedes may have thought of the case of the pyramid as being the more fundamental and as really involving that of the cone. Or perhaps 'figure' may be intended for 'type of figure'.

‡ Cf. Preface to *Quadrature of Parabola*.

Archimedes then says that he first applied his method to the quadrature of the parabola. He had already explained this quadrature in his treatise on the subject, whose preface runs as follows [6]:

'Archimedes to Dositheus greeting.

When I heard that Conon, who was my friend in his lifetime, was dead, but that you were acquainted with Conon and withal versed in geometry, while I grieved for the loss not only of a friend but of an admirable mathematician, I set myself the task of communicating to you, as I had intended to send to Conon, a certain geometrical theorem which had not been investigated before but has now been investigated by me, and which I first discovered by means of mechanics and then exhibited by means of geometry. Now some of the earlier geometers tried to prove it possible to find a rectilineal area equal to a given circle and a given segment of a circle; and after that they endeavoured to square the area bounded by the section of the whole cone* and a straight line, assuming lemmas not easily conceded, so that it was recognised by most people that the problem was not solved. But I am not aware that any one of my predecessors has attempted to square the segment bounded by a straight line and a section of a right-angled cone [a parabola], of which problem I have now discovered the solution. For it is here shown that every segment bounded by a straight line and a section of a right-angled cone [a parabola] is four-thirds of the triangle which has the same base and equal height with the segment, and for the demonstration of this property the following lemma is assumed: that the excess by which the greater of (two) unequal areas exceeds the less can by being added to itself, be made to exceed any given finite area.† The earlier geometers have also used this lemma; for it is by the use of this same lemma that they have shown that circles are to one another in the duplicate ratio of their diameters, and that spheres are to one another in the triplicate ratio of their diameters,‡ and further that every pyramid is one third part of the prism which has the same base with the pyramid and equal height; also, that every cone is one third part of the cylinder having the same base as the cone and equal height they proved by assuming a certain lemma similar to that aforesaid. And, in the result, each of the aforesaid theorems has been accepted no less than those proved without the lemma. As therefore my work now published has satisfied the same test as the propositions referred to, I have written out the proof and send it to you, first as investigated by means of mechanics, and afterwards too as demonstrated by geometry. Prefixed are, also, the elementary propositions in conics which are of service in the proof (στοιχεῖα κωνικὰ χρείαν ἔχοντα ἐς τὰν ἀπόδειξιν). Farewell.

---

* Probably the ellipse.

† Compare Definition 4 of Book V of Euclid's *Elements*.

‡ The last proposition of Book XII of Euclid's *Elements*. The passage in fact summarizes all the propositions of the book.

We shall not misrepresent Archimedes's work if we give a modern proof of those properties of the parabola which must be understood if we are to follow his reasoning. The Greek treatment of conic sections by use of geometrical algebra* is in fact identical with the methods of modern analytical geometry.

A parabola may be represented by a trinomial of the second degree.

A parabola passing through the origin, C (Fig. 4.15), has an equation

$$y = ax^2 - bx.$$

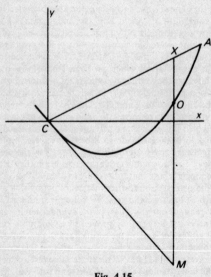

Fig. 4.15

A point A on the curve which has abscissa x_0 has coordinates

$$x_0 \quad \text{and} \quad ax_0^2 - bx_0;$$

---

* The expression comes from Zeuthen.

and the straight line CA has gradient

$$y_0 : x_0 = ax_0 - b.$$

The tangent to the parabola at the origin is the straight line $y = -bx$.

Let X, O and M be points on the straight line CA, the parabola and the tangent at C, each having the same abscissa x. Then their ordinates are

For X: $\quad x(ax_0 - b)$ so $\overline{OX} = ax(x_0 - x)$

For O: $\quad x(ax - b)$ and $\overline{MX} = ax\, x_0$.

For M: $\quad -bx$

Therefore

$$\frac{\overline{MX}}{\overline{OX}} = \frac{ax\, x_0}{ax(x_0 - x)} = \frac{x_0}{x_0 - x} = \frac{\overline{CA}}{\overline{XA}}.$$

Archimedes uses this result in his mechanical method. Now let D be the mid-point of the chord CA (Fig. 4.16). Its coordinates are

$$\left(\frac{x_0}{2}, \frac{x_0}{2}(ax_0 - b) \right).$$

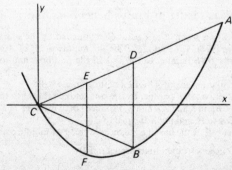

Fig. 4.16

Let B be the point on the parabola which has the same abscissa as D. Its ordinate is

$$a\frac{x_0^2}{4} - b\frac{x_0}{2}, \text{ so } \overline{BD} = a\frac{x_0^2}{4}.$$

The tangent to the parabola at B has gradient

$$y' \equiv 2ax - b = ax_0 - b,$$

which is the same as the gradient of the chord CA. The tangent is thus parallel to the chord.

Archimedes therefore calls B the *vertex* of the segment CBA.

Let E be the mid-point of CD. It has abscissa $x_0/4$. The line EF parallel to the axis of the parabola and passing through the mid-point of CD, also passes through the mid-point of CB. The point F, where this line intersects the parabola, is therefore the *vertex* of the segment CFB.

The ordinate of E is $(x_0/4)(ax_0 - b)$, and that of F is $a(x_0/4)^2 - b(x_0/4)$, so $\overline{FE} = \frac{3}{16}ax_0^2$.

Therefore $FE = \frac{3}{4}DB$.

Archimedes uses this result in his geometrical method.

MECHANICAL METHOD

Archimedes's letter to Eratosthenes continues:

Let ABC be a segment of a parabola* bounded by the straight line AC and the parabola ABC, and let D be the middle point of AC. Draw the straight line DBE parallel to the axis of the parabola and join AB, BC [Fig. 4.17].

Then shall the segment ABC be $\frac{4}{3}$ of the triangle ABC.

From A draw AKF parallel to DE, and let the tangent to the parabola at C meet DBE in E and AKF in F. Produce CB to meet AF in K, and again produce CK to H, making KH equal to CK.

Consider CH as the bar of a balance, K being its middle point.

Let MO be any straight line parallel to ED.

---

* For the sake of simplicity we use the modern word.

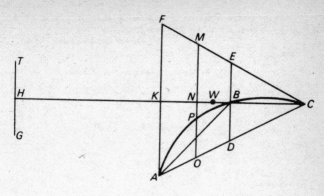

Fig. 4.17

If *MO* cuts the line *HKC* in *N* and the parabola in *P* we have (by the proposition proved on p. 120 above)

$$MO:OP = CA:AO = CK:KN = KH:KN.$$

Take a straight line *TG* equal to *OP*, and place it with its centre of gravity at *H*, so that *TH = HG*; then, since *N* is the centre of gravity of the straight line *MO*, and

$$MO:TG = HK:KN,$$

it follows that *TG* at *H* and *MO* at *N* will be in equilibrium about *K*.

[*On the Equilibrium of Planes*, I. 6, 7]

Similarly, for all other straight lines parallel to *DE* and meeting the arc of the parabola, (1) the portion intercepted between *FC*, *AC* with its middle point on *KC* and (2) a length equal to the intercept between the curve and *AC* placed with its centre of gravity at *H* will be in equilibrium about *K*.

Therefore *K* is the centre of gravity of the whole system consisting (1) of all the straight lines as *MO* intercepted between *FC*, *AC* and placed as they actually are in the figure and (2) of all the straight lines placed at *H* equal to the straight lines as *PO* intercepted between the curve and *AC*.

And, since the triangle *CFA* is made up of all the parallel lines like *MO*, and the segment *CBA* is made up of all the straight lines like *PO* within the curve,

it follows that the triangle, placed where it is in the figure, is in equilibrium about *K* with the segment *CBA* placed with its centre of gravity at *H*.

Divide *KC* at *W* so that *CK = 3KW* ;

then W is the centre of gravity of the triangle ACF; 'for this is proved in the books on equilibrium' ($\dot{\epsilon}\nu$ $\tau o\hat{\iota}\varsigma$ $\dot{\iota}\sigma o\rho\rho o\pi\iota\kappa o\hat{\iota}\varsigma$).

[Cf. *On the Equilibrium of Planes* I. 15]

Therefore

$$\triangle ACF : (\text{segment } ABC) = HK : KW$$
$$= 3 : 1.$$

Therefore

$$\text{segment } ABC = \tfrac{1}{3}\triangle ACF.$$

But

$$\triangle ACF = 4\triangle ABC.$$

Therefore

$$\text{segment } ABC = \tfrac{4}{3}\triangle ABC.$$

Since this proof seems to him to be less than completely rigorous, Archimedes takes it up again in *The Quadrature of the Parabola* and replaces the infinite number of lines parallel to the diameter by trapezia. He thus shows that the triangle CAF is neither smaller nor greater than three times the segment.

However, since he still had to appeal to mechanics, this procedure did not altogether satisfy him. In the second part of his treatise on the quadrature of the parabola he bases his work on Eudoxus's method of cubing the pyramid and inscribes the triangle ABC in the segment ABC with vertex B at the vertex of the segment (Fig. 4.18).

He shows that this triangle has an area greater than half that of the segment by comparing it with the parallelogram with base AC and upper boundary formed by the tangent to the parabola at the point B.

The segment consists of the triangle and two small segments, such as AFB. In each of these he inscribes a triangle such as AFB. Now,

$$\triangle AFB : \triangle ABD = FG : DB.$$

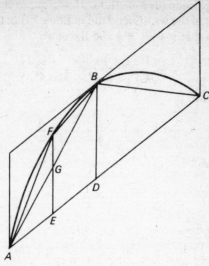

Fig. 4.18

But

$$GE = \tfrac{1}{2}BD \text{ and } FE = \tfrac{3}{4}BD$$

$$\therefore \quad FG:BD = \tfrac{1}{4}.$$

$$\therefore \quad \triangle AFB = \tfrac{1}{4}\triangle ABD$$

i.e.

$$\triangle AFB = \tfrac{1}{8}\triangle ABC.$$

The two new inscribed triangles are therefore, together, equal to $\tfrac{1}{4}\triangle ABC$. Let us take the area of this last triangle as unity. By continuing the above process we obtain the sums

$$1; \quad 1 + \tfrac{1}{4}; \quad 1 + \tfrac{1}{4} + \tfrac{1}{16}; \quad 1 + \tfrac{1}{4} + \tfrac{1}{16} + \tfrac{1}{64} \text{ etc.} \ldots$$

which measure areas closer and closer to the area we require, and a sum could be found whose difference from this area was less than any given quantity.

This is Eudoxus's method. However, Archimedes does not use it in the way Euclid did in Book XII of the *Elements*. He notes that $\frac{1}{4} + \frac{1}{12} = \frac{1}{3}$ and therefore

$$\frac{1}{4^n} + \frac{1}{3 \times 4^n} = \frac{1}{3 \times 4^{n-1}}.$$

To the sum

$$S_n = 1 + \frac{1}{4} + \cdots + \frac{1}{4^n}$$

let us add

$$\frac{1}{3} \cdot \frac{1}{4^n}.$$

We obtain

$$S_n + \frac{1}{3} \cdot \frac{1}{4^n} = 1 + \frac{1}{4} + \cdots + \frac{1}{4^{n-1}} + \frac{1}{3 \times 4^{n-1}}$$

$$= 1 + \frac{1}{4} + \cdots + \frac{1}{4^{n-2}} + \frac{1}{3 \times 4^{n-2}} \quad \text{etc.}$$

$$= \frac{4}{3}.$$

We can now use the method of *reductio ad absurdum* to show that the segment of the parabola has measure $\frac{4}{3}$, i.e. is $\frac{4}{3}$ times the area of the inscribed triangle.

For, if the proposition is untrue, the area of segment is either greater or less than $\frac{4}{3}$ times the inscribed triangle. Let it be greater. We shall denote it by Σ and write $\Sigma = \frac{4}{3} + U$. We inscribe in the segment a series of triangles such that

$$\Sigma - S_n < U,$$

which is always possible. Then $\Sigma < S_n + U$. But $S_n < \frac{4}{3}$, so $S < \frac{4}{3} + U$, which is a contradiction.

We therefore suppose that Σ is less than $\frac{4}{3}$, and write $\Sigma = \frac{4}{3} - V$.

Let us inscribe in the segment a series of triangles such that the last term added, $1/4^n$, is less than V. Thus, since

$$\Sigma + V = \tfrac{4}{3} \quad \text{and} \quad S_n < \Sigma, S_n + V < \tfrac{4}{3},$$

whence

$$\frac{4}{3} - \frac{1}{3} \cdot \frac{1}{4^n} + V < \frac{4}{3},$$

i.e.

$$V < \frac{1}{3} \cdot \frac{1}{4^n},$$

which is a contradiction.

The segment therefore must be $\frac{4}{3}$ of the triangle.

The success of his mechanical method was to lead Archimedes to new discoveries. He writes to Eratosthenes:

We can investigate by the same method the propositions that

(1) *Any sphere is (in respect of solid content) four times the cone with base equal to a great circle of the sphere and height equal to its radius;* and

(2) *the cylinder with base equal to a great circle of the sphere and height equal to the diameter is* $1\frac{1}{2}$ *times the sphere.*

(1) Let $ABCD$ be a great circle of a sphere, and AC, BD diameters at right angles to one another [Fig. 4.19].

Let a circle be drawn about BD as diameter and in a plane perpendicular to AC, and on this circle as base let a cone be described with A as vertex. Let the surface of this cone be produced and then cut by a plane through C parallel to its base; the section will be a circle on EF as diameter. On this circle as base let a cylinder be erected with height and axis AC, and produce CA to H, making AH equal to CA.

Let CH be regarded as the bar of a balance, A being its middle point.

Draw any straight line MN in the plane of the circle $ABCD$ and parallel to BD. Let MN meet the circle in O, P, the diameter AC in S, and the straight lines AE, AF in Q, R respectively.

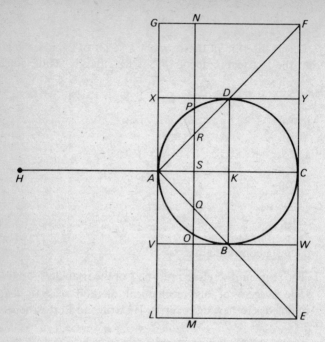

Fig. 4.19

Through MN draw a plane at right angles to AC;
this plane will cut the cylinder in a circle with diameter MN, the sphere in a
circle with diameter OP, and the cone in a circle with diameter QR.

Now, since

$$MS = AC, \quad \text{and} \quad QS = AS$$
$$MS \cdot SQ = CA \cdot AS$$
$$= AO^2$$
$$= OS^2 + SQ^2.$$

And, since

$$HA = AC,$$

$$HA:AS = CA:AS$$

$$= MS:SQ$$

$$= MS^2:MS.SQ$$

$$= MS^2:(OS^2 + SQ^2), \text{ from above,}$$

$$= MN^2:(OP^2 + QR^2)$$

$$= \text{(circle, diam. } MN)\text{:(circle, diam. } OP + \text{circle, diam. } QR).$$

That is,

$$HA:AS = \text{(circle in cylinder):(circle in sphere } \dotplus \text{ circle in cone).}$$

Therefore the circle in the cylinder, placed where it is, is in equilibrium, about A, with the circle in the sphere together with the circle in the cone, if both the latter circles are placed with their centres of gravity at H.

Similarly for the three corresponding sections made by a plane perpendicular to AC and passing through any other straight line in the parallelogram LF parallel to EF.

If we deal in the same way with all the sets of three circles in which planes perpendicular to AC cut the cylinder, the sphere and the cone, and which make up those solids respectively, it follows that the cylinder, in the place where it is, will be in equilibrium about A with the sphere and the cone together, when both are placed with their centres of gravity at H.

Therefore, since K is the centre of gravity of the cylinder,

$$HA:AK = \text{(cylinder):(sphere } \dotplus \text{ cone } AEF).$$

But $HA = 2AK$;

therefore

$$\text{cylinder} = 2 \text{ (sphere } + \text{ cone } AEF).$$

Now

$$\text{cylinder} = 3 \text{ (cone } AEF);$$

therefore

$$\text{cone } AEF = 2 \text{ (sphere).}$$

But, since $EF = 2BD$,

$$\text{cone } AEF = 8 \text{ (cone } ABD);$$

therefore

$$\text{sphere} = 4 \text{ (cone } ABD).$$

(2) Through B, D draw VBW, XDY parallel to AC;
and imagine a cylinder which has AC for axis and the circles on VX, WY as diameters for bases.

Then

$$\text{cylinder } VY = 2 \text{ (cylinder } VD)$$
$$= 6 \text{ (cone } ABD)$$
$$= \tfrac{3}{2} \text{ (sphere), from above.}$$

Q.E.D.

'From this theorem, to the effect that a sphere is four times as great as the cone with a great circle of the sphere as base and with height equal to the radius of the sphere, I conceived the notion that the surface of any sphere is four times as great as a great circle in it; for, judging from the fact that any circle is equal to a triangle with base equal to the circumference and height equal to the radius of the circle, I apprehended that, in like manner, any sphere is equal to a cone with base equal to the surface of the sphere and height equal to the radius.'

The matter is taken up again in the treatise *On the sphere and cylinder*, and rigorous proofs are given. This work is addressed to Dositheus, and the preface begins:

Archimedes to Dositheus greeting.

On a former occasion I sent you the investigations which I had up to that time completed, including the proofs, showing that any segment bounded by a straight line and a section of a right-angled cone [a parabola] is four-thirds of the triangle which has the same base with the segment and equal height. Since then certain theorems not hitherto demonstrated (ἀνελέγκτων) have occurred to me, and I have worked out the proofs of them. They are these: first, that the surface of any sphere is four times its greatest circle (τοῦ μεγίστου κύκλου); next, that the surface of any segment of a sphere is equal to a circle whose radius (ἡ ἐκ τοῦ κέντρου) is equal to the straight line drawn from the vertex (κορυφή) of the segment to the circumference of the circle which is the base of the segment; and, further, that any cylinder having its base equal to the greatest circle of those in the sphere, and height equal to the diameter of the sphere, is itself [*i.e.* in content] half as large again as the sphere, and its surface also [including its bases] is half as large again as the surface of the sphere. Now these properties were all along naturally inherent in the figures referred to (αὐτῇ τῇ φύσει προϋπῆρχεν περὶ τὰ εἰρημένα σχήματα), but remained unknown to those who were before my time engaged in the study of geometry.

Archimedes gives six definitions, which he calls Axioms (ἀξιώματα):

1. There are in a plane certain terminated bent lines (καμπύλαι γραμμαὶ πεπερασμέναι), which either lie wholly on the same side of the straight lines joining their extremities, or have no part of them on the other side.

2. I apply the term **concave in the same direction** to a line such that, if any two points on it are taken, either all the straight lines connecting the points fall on the same side of the line, or some fall on one and the same side while others fall on the line itself, but none on the other side.

3. Similarly also there are certain terminated surfaces, not themselves being in a plane but having their extremities in a plane, and such that they will either be wholly on the same side of the plane containing their extremities, or have no part of them on the other side.

4. I apply the term **concave in the same direction** to surfaces such that, if any two points on them are taken, the straight lines connecting the points either all fall on the same side of the surface, or some fall on one and the same side of it while some fall upon it, but none on the other side.

5. The **solid sector** is the figure enclosed between the surface of a sphere and a cone having the same centre.

6. The **solid rhombus** is the figure formed by two cones having the same axis and the same base and vertices on opposite sides of the base.

Then come five *postulates* or demands, which Archimedes calls Λαμβανόμενα which are different in several respects from those to be found in Euclid's *Elements*:

1. Of all lines which have the same extremities the straight line is the least.*

2. Of other lines in a plane and having the same extremities, [any two] such are unequal whenever both are concave in the same direction and one of them is either wholly included between the other and the straight line which has the same extremities with it, or is partly included by, and is partly common with, the other; and that [line] which is included is the lesser [of the two].

3. Similarly, of surfaces which have the same extremities, if those extremities are in a plane, the plane is the least [in area].

4. Of other surfaces with the same extremities, the extremities being in a plane, [any two] such are unequal whenever both are concave in the same direction and one surface is either wholly included between the other and the plane which has the same extremities with it, or is partly included by, and partly common with, the other; and that [surface] which is included is the lesser [of the two in area].

5. Further, of unequal lines, unequal surfaces, and unequal solids, the greater exceeds the less by such a magnitude as, when added to itself, can be made to exceed any assigned magnitude among those which are comparable with [it and with] one another.

---

* This property of the straight line is a *postulate* in the work of Archimedes, but is taken as a *definition* in Campanus's Latin edition of the *Elements* in the thirteenth century, and by many later authors. Until the end of the nineteenth century it was the definition generally adopted.

From these definitions and postulates Archimedes proceeds to deduce rigorously the results suggested to him by his mechanical method and by his mathematical intuition.

We shall not work through all the forty-four propositions of this treatise, which is a mathematical classic of the same calibre as Euclid's *Elements*. It contains such well-known results as the formulae for the volume and surface area of a sphere, for the area of a zone of a sphere and for the volume of a sector of a sphere.

To give some idea of the ingenious and rigorous nature of Archimedes's methods let us examine Proposition 9:

Consider a cone of revolution.

The triangle *DAC* is formed by two generators of the cone, *DA* and *DC*, and a chord *AC* of its circular base (Fig. 4.20).

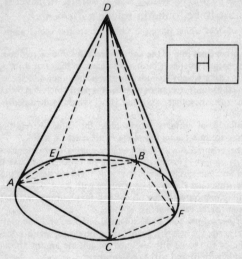

Fig. 4.20

We wish to show that the area of this triangle is less than the area of the nappe of the cone bounded by the generators *DA* and *DC*.

Archimedes proceeds as follows: He first states that in any pyramid $DABC$, with equal edges DA, DB and DC one of the faces passing through the vertex is always less than the sum of the two others.

This proposition is not proved in any known Ancient text. It is, however, quite easy to verify, and we shall accept it.

Let B be the mid-point of the arc ABC which forms the base of the nappe of the cone.

Since $\triangle ADB + \triangle CDB > \triangle ADC$ let us write $\triangle ADB + \triangle CDB = \triangle ADC + H$ where H is some area.

If the segments AEB and BFC are together less than H, then Postulate 4, above, gives:

$$\text{segment } AEB + \text{nappe } DAEB > \triangle ADB$$

and

$$\text{segment } BFC + \text{nappe } DBFC > \triangle BCD.$$

Hence, adding the terms and using the fact that H is greater than the sum of the segments, we have

$$\text{H} + \text{nappe } DABC > \triangle ADB + \triangle BDC$$

or

$$\text{H} + \text{nappe} > \triangle ADC + \text{H}$$

and nappe $> \triangle ADC$.

If H is less than the combined segments of the circle, we take the mid-points E and F of these segments, and repeat the process until the segments between the circle and the line $AEBFC$ are less than H, which is always possible.

Then, from Axiom 1,

$$\text{segments} + \text{nappe} > \text{sum of } \triangle \text{s } AED + \cdots + DFC.$$

This sum is itself greater than the sum of the two triangles ABD and CBD.

Hence H + nappe > $\triangle ADC$ + H and finally nappe > $\triangle ADC$.

4.3. Apollonius

Reliable accounts of the life of the third of the great geometers (i.e. the third chronologically, we have not attempted to judge their relative merits) are almost as scarce as in the case of Euclid. Historians made him either 25 or 40 years younger than Archimedes. From a papyrus found at Herculaneum he is presumed to have been active around 170 BC.

He was born at Perga in Pamphylia, and, as he himself tells us in the various prefaces to his *Conics*, he lived in Alexandria and at Pergamum.

His chief work is his great treatise on the conic sections. In Chapter 3, Vol. 2, we shall describe the principle of the method he uses in that work. The method is much closer to modern analytical geometry, which in fact derives from the work of Apollonius, *via* Vieta, Descartes and Fermat, than it is to the methods of pure geometry.

The first four books, the only ones to have come down to us in the original Greek, must be taken to represent the theory of conics as developed by Apollonius's predecessors: Menaechmus, Euclid, Aristaeus, Archimedes and Conon, rather than as Apollonius's own work. Nevertheless he has generalized their theory, principally by considering a completely general plane section of a right or oblique cone with a circular base. He invented the modern names for the conic sections*: ellipse, parabola and hyperbola, although he uses this last term only in reference to one of the two branches of the curve, denoting the whole curve by the term 'opposite sections'.

The last four books, unlike the first four, deal with the author's original work. The fifth considers normals to

---

* See Vol. 2, p. 92.

conics and determines their envelopes. The sixth deals with equality and similarity between conics. We should note that Apollonius, unlike Euclid, here uses 'equality' in the sense of the modern term 'congruence'.

Book VI. Definition 1:

> Conic sections are said to be **equal** when one can be applied to the other in such a way that they everywhere coincide and nowhere cut one another. When this is not the case they are **unequal**.

In the seventh book we find Apollonius's two theorems on conjugate diameters. These last three books have come down to us only in an Arabic translation. The eighth book is lost.

The general preface to the treatise reads as follows [7]:

> Apollonius to Endemus, greeting.
>
> If you are in good health and things are in other respects as you wish, it is well; with me too things are moderately well. During the time I spent with you at Pergamum I observed your eagerness to become acquainted with my work in conics; I am therefore sending you the first book, which I have corrected, and I will forward the remaining books when I have finished them to my satisfaction. I dare say you have not forgotten my telling you that I undertook the investigation of this subject at the request of Naucrates the geometer, at the time when he came to Alexandria and stayed with me, and, when I had worked it out in eight books, I gave them to him at once, too hurriedly, because he was on the point of sailing; they had therefore not been thoroughly revised, indeed I had put down everything just as it occurred to me, postponing revision till the end. Accordingly I now publish, as opportunities serve from time to time, instalments of the work as they are corrected. In the meantime it has happened that some other persons also, among those whom I have met, have got the first and second books before they were corrected; do not be surprised therefore if you come across them in a different shape.
>
> Now of the eight books the first four form an elementary introduction. The first contains the modes of producing the three sections and the opposite branches (of the hyperbola), and the fundamental properties subsisting in them, worked out more fully and generally than in the writings of others. The second book contains the properties of the diameters and the axes of the sections as well as the asymptotes, with other things generally and necessarily used for determining limits of possibility ($\delta\iota o\rho\iota\sigma\mu o i$);* and

---

* That is, the conditions required for problems to be soluble.

what I mean by diameters and axes respectively you will learn from this book. The third book contains many remarkable theorems useful for the syntheses of solid loci* and for *diorismi*; the most and prettiest of these theorems are new, and it was their discovery which made me aware that Euclid did not work out the synthesis of the locus with respect to three and four lines,† but only a chance portion of it, and that not successfully; for it was not possible for the said synthesis to be completed without the aid of the additional theorems disdovered by me. The fourth book shows in how many ways the sections of cones can meet one another and the circumference of a circle; it contains other things in addition, none of which have been discussed by earlier writers, namely the questions in how many points a section of a cone or a circumference of a circle can meet [a double-branch hyperbola, or two double-branch hyperbolas can meet one another].

The rest of the books are more by way of surplusage ($\pi\epsilon\rho\iota\sigma\nu\sigma\iota\alpha\sigma\tau\iota\kappa\acute{\omega}$-$\tau\epsilon\rho\alpha$): one of them deals somewhat fully with *minima* and *maxima*‡, another with equal and similar sections of cones, another with theorems of the nature of determinations of limits, and the last with determinate conic problems§. But of course, when all of them are published, it will be open to all who read them to form their own judgement about them, according to their own individual tastes. Farewell.

Most of Apollonius's works have not come down to us. His treatise *De sectione rationis*, in two books, was translated from Arabic into Latin by Halley 1703. It deals with the problem of drawing through a given point, a straight line to cut off segments of two given lines, the segments, measured from given points, to be in a given ratio.

---

* Geometrical loci which are conics.

† Pappus says: If three straight lines be given in position, and from one and the same point straight lines be drawn to meet the three straight lines at given angles, and if the ratio of the rectangle contained by two of the straight lines towards the square on the remaining straight line be given, then the point will lie on a solid locus given in position, that is on one of the three conic sections. And if straight lines be drawn to meet at given angles four straight lines given in position, and the ratio of the rectangle contained by two of the straight lines so drawn towards the rectangle contained by the remaining two be given, then in the same way the point will lie on a conic section given in position. (*Mathematical Collection.* Preface to Book VII, translated by Thomas.)

‡ Book V: the normal to a conic from a given point is the shortest or the longest line that can be drawn from that point to the curve.

§ That is, problems which have a finite number of solutions.

Apollonius's lost works are known to us only from Pappus's commentaries on them.

The two books entitled *De spatii sectione* deal with a problem like that of *De sectione rationis* but the segments instead of being in a given ratio are required to form a given rectangle, that is to have a given product.

Willebrord Snell attempted to reconstruct these books in 1607.

The two books of *De sectione determinata* refer to a theory whose place in ancient mathematics was rather like that of inversion in modern geometry. A reconstruction of this work, by Willebrord Snell, was published in 1608, under the title *Appollonius Batavus*. Pappus summarises the contents as follows:

> To cut a given infinite straight line in a point so that the intercepts between this point and given points on the line shall furnish a given ratio, the ratio being that of the square on one intercept, or the rectangle contained by two, towards the square on the remaining intercept, or the rectangle contained by the remaining intercept and a given independent straight line, or the rectangle contained by two remaining intercepts, whichever way the given points [are situated]. . . .

Two books *On Contacts* (ἐπαφαί) were reconstructed by Vieta in 1600, under the title *Apollonius Gallus*. Pappus summarizes them as follows:

> Given three entities, of which any one may be a point or a straight line or a circle, to draw a circle which shall pass through each of the given points, so far as it is points which are given, or to touch each of the given lines.

The page of *Apollonius Gallus* shown in Plate VII considers the problem: Given two circles, to draw through a given point a circle which touches both of them. The centres of the two circles are shown as K and L. The given point is I. Vieta begins by finding the point, M, the centre of similitude of the two circles. He calls it 'the point such that the lines from it cut similar segments from the two circles'. He

APOLLONIVS

Cadant enim ex centris K, L in subtensas BC FG perpendiculares KR LS, & connectantur semidiametri KC LG. Quoniam BC FG proponuntur similia suorum circulorum segmenta, RC verò & SG sunt semisses subtensarum BC FG, ideò similia sunt triangula KRC LSG & parallelæ KC LG, atque adeò anguli CKD GLH similes ac denique subtensæ eorum amplitudini CD GH parallelæ. Quare est vt MD ad MC ita MH ad MG. Sed MD ad MC est vt MB ad MA. Et ideò quod fit sub MB MG ei quod fit sub MA MH æquale.

Jisdem positis, Aio eadem ratione id quod fit sub ME MD æquari ei quod fit sub MF MC.

Quoniam enim EFGH est in circulo ideò est MH ad MG sicut MF ad ME. Sed MH ad MG est sicut MD ad MC. Ergo est MF ad ME sicut MD ad MC, & ideò quod fit sub ME MD ei quod fit sub MF MC est æquale.

PROBLEMA IX.

Datis duobus circulis, & puncto, per datum punctum circulum describere quem duo dati circuli contingant.

Sint dati duo circuli, vnus ABCD alter EFGH, & præterea datum I punctum. Oportet per I punctum circulum describere quem circuli ABCD, EFGH contingant. Circulorum ABCD EFGH iungantur centra K, L & in KL inueniatur ex antecedente, quod primo loco præmissum est, lemmate, M punctum, à quo, cum ducentur rectæ lineæ secantes circulos ABCD EFGH, similia sint segmenta. Ipsa verò kL secet ex diametro circulum ABCD in signis A, D, circulum verò EFGH in signis E, H, & ita secetur MI in N, vt quod fit sub MI MN æquale sit ei quod fit sub MH MA, & per I, N puncta describatur circulus qui à circulo ABCD tangatur. Id enim iam docuit Problema octauum. Et sit contactus in B, & connectatur BM secans circulum ABCD in B, C, circulum verò EFGH in F, G. Quod fit igitur sub MG MB æquale est ei quod fit sub MH MA, id est ex constructione ei quod fit sub MN MI ex antecedente, quod secundo

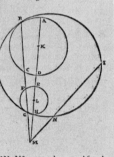

B.N.

PLATE VII. From Vieta, *Apollonius Gallus* (1600).

finds the point N on MI such that $MN \times MI = MH \times MA$ and thus reduces the problem to that of constructing a circle through I and N which touches the first of the two given circles. This problem has already been solved.

These problems of circles in contact with one another interested French geometers of the nineteenth century, particularly Gergonne, Poncelet and Gaultier de Tours. They are one of the sources of modern synthetic geometry.

According to Pappus, Apollonius's two books *On Inclinations* dealt with the problem of constructing a straight line through a given point such that two given lines cut off from it a segment of given length. Marino Ghetaldi of Ragusa published a reconstruction of this work in 1607, under the title *Apollonius Redivivus*.

Two books of *Plane Loci* were reconstructed by Fermat around 1630, by Frans van Schooten in 1656 and by Robert Simson in 1749.

Here is what Pappus says of this last treatise:

Loci in general are termed *fixed*, as when Apollonius at the beginning of his own *Elements* says the locus of a point is a point, the locus of a line is a line, the locus of a surface is a surface and the locus of a solid is a solid; or *progressive*, as when it is said that the locus of a point is a line, the locus of a line is a surface and the locus of a surface is a solid; or *circumambient* as when it is said that the locus of a point is a surface and the locus of a line is a solid. [... the loci described by Eratosthenes as having reference to means belong to one of the aforesaid classes, but from a peculiarity in the assumptions are unlike them.]

The ancients had regard to the arrangement of these plane loci with a view to instruction in the elements; heedless of this consideration, their successors have added others, as though the number could not be infinitely increased if one were to make additions from outside that arrangement. Accordingly I shall set out the additions later, giving first those in the arrangement, and including them in this single enunciation:

If two straight lines be drawn, from one given point or from two, which are in a straight line or parallel or include a given angle, and either bear a given ratio one towards the other or contain a given rectangle, then, if the locus of the extremity of one of the lines be a plane locus given in position, the locus of the extremity of the other will also be a plane locus given in position, which will sometimes be of the same kind as the former, sometimes of a different kind, and will sometimes be similarly situated with respect to the *straight line, sometimes contrariwise.* These different cases arise according to the differences in the suppositions.* ...

The other propositions are as follows:

If a straight line be drawn of a given length and parallel to a fixed straight line and if the locus of one of its extremities be a fixed straight line then the locus of its other extremity will also be a fixed straight line.

If straight lines be drawn from a point to meet two fixed straight lines, which may or may not be parallel to one another, and if these lines cut them at given angles and either bear a given ratio to one another or are such that the sum of the length of one and a length which bears a fixed ratio to the other is given, then the locus of the point will be a fixed straight line.

If from some point lines be drawn at given angles to cut fixed parallel lines at given angles, and if they cut off from these parallel lines segments, each measured from some fixed point on the parallel lines, such that these segments bear a given ratio one towards the other or contain a given area or are such that the sum or difference of the areas constructed on the lines that have been drawn is equivalent to a given area, then the locus of the point will be a fixed straight line.

The second book contains the following:

If straight lines drawn from two points meet at a point, and if the difference of the squares constructed on these lines is equal to some given area, the locus of the meet of the two lines is a fixed straight line.

On the other hand, if these straight lines bear a given ratio one towards the other, the locus of their meet will be either a straight line or the circumference of a circle.†

Given a fixed straight line and a fixed point on it, if a straight segment be drawn from this point, and if from the extremity of the segment a perpendicular be dropped to the fixed straight line, and if the square of the segment is equivalent to the rectangle contained by a given straight line and the segment cut off by the perpendicular, measured from the given point from another given point on the fixed straight line, then the locus of the extremity of the first segment will be the circumference of a fixed circle.

If the straight lines drawn from two given points meet at a point, and if the square on the one is greater than the square on the other in a given ratio by a given constant‡, then the locus of the meet of the two lines is the circumference of a fixed circle.

If straight lines from any number of points all meet at one point, and if the figures described on all these straight lines are equivalent to a given area, the locus of the meet of the lines will be the circumference of a fixed circle.§

---

* We are concerned here with what would, in modern terms, be called the transformation of a circle or a straight line by inversion.

† This locus was known to Aristotle. See *Meteora.*

‡ y is greater by some given number than x in a certain ratio if $y = px + q$.

§ If x, y and z are the lengths of these lines we have $px^2 + qy^2 + rz^2 = a^2$.

If the straight lines drawn from two points meet at a point, and if the straight line drawn from this point parallel to a given straight line cuts off from a fixed straight line a segment, measured from a given point of the line, and if the sum of the figures described on the meeting straight lines is equivalent to the rectangle contained by a given straight line and the segment cut off from the fixed straight line, the locus of the meet of the two straight lines is the circumference of a fixed circle.*

Given a point inside a fixed circle, if a straight line be drawn through that point, and if a point be taken on this line outside [the circle], and if the square constructed on the line from this point to the given interior point, either the square alone or the square plus the rectangle contained by the two interior segments, if this is equivalent to the rectangle contained by the whole of this straight line and the segment of it cut off outside the circle, the locus of the exterior point will be a fixed straight line . . . .

---

* A and B are fixed. There is a line Cx through C. We consider the segments $AM = u$, $BM = v$. A line My is drawn through M, parallel to a fixed direction. My cuts Cx in P. We have

$$au^2 + bv^2 = l \times CP.$$

5. Autumn and Winter of Science

*Mais où sont les neiges d'antan?**
  François Villon (1430?–1466?)

5.1. Tradition

Between the Hellenistic period and the rebirth of mathematics in the West, there were several centuries during which the historical contribution even of original thinkers can nevertheless best be seen as the transmission of traditional knowledge.

This tradition was modified in some respects, and points of view also changed, but evolution on the whole was slow, and the only sharp and important change was the first appearance of the decimal point, in Hindu work, in about the fifth century AD.

It would not be possible to list all the mathematicians of these fifteen hundred years, so we shall content ourselves with mentioning some of the better known figures.

5.2. The Greeks

We shall begin with four Greek scholars who made contributions to trigonometry.

Theodosius, long thought to have been a native of Tripoli, and later believed to have been born in Bithynia, seems to have lived earlier than the first century AD. His name is

---

* But where are the snows of yesteryear? Trans. D. G. Rossetti.

associated with a treatise on the sphere, which we shall discuss briefly in Chapter 6, Vol. 2.

No mathematical works have come down to us from the greatest astronomer of Antiquity, Hipparchus. He was born at Nicaea, in Bithynia. His astronomical observations were made at Rhodes and Alexandria from 161 to 126 BC.

Menelaus, whose treatise on the sphere has survived, is known to have made astronomical observations in the time of the emperor Trajan, at the end of the first century AD.

Claudius Ptolemy of Alexandria, famous as the author of the *Almagest*, made observations from 125 until 141 AD and must have written his great work during the reign of Antoninus Pius.

Bailly, who became Mayor of Paris and died on the scaffold, wrote a *History of Modern Astronomy*, published in 1785, in which he says of Ptolemy:

'Ptolemy was born at Ptolemais in Egypt. The similarity between the names, without further evidence, has led some to conclude that he was of royal descent; but his genius was such that it did not need this aid of a royal name, often of little value, in order to pass his fame on to posterity. He worked during the reigns of Hadrian and Antoninus; his observations, which are the proof of an astronomer's activity, cover an interval of fourteen years. He observed an eclipse of the moon in the ninth year of Hadrian's reign, that is in AD 125; he observed some stars in the second year of Antoninus's reign, that is in AD 139. This then is the period of his active work. . . .

'In addition to his own work, this astronomer had the merit of collecting together the work of others and shaping it into a corpus of knowledge whose unity and usefulness have resisted the ravages of time. His *Almagest* is the link between ancient and modern astronomy; it is rather like one of those trading posts: centres of commerce which receive goods from one part of the world to transmit them to another. It preserves observations which are important because of their early date, since without them we should not know the mean movements of the planets as accurately as they were known to Hipparchus and Ptolemy. The work, moreover, contains descriptions of methods or the origins of methods which are still in use today. For a long time it was the elementary textbook used all over the world,* and all credit is due to the author.'

Along with these astronomers, our survey of the history of mathematics must also include an engineer, Heron of Alexandria. We know practically nothing about his life, but his work is of considerable importance.

He is now usually believed to have lived in the first century AD, though this is still a matter of controversy.

The list of his works is a long one, made even longer by apocryphal additions, some of which date from as late as 1000 AD.

The principal works are:

1. THE METRICA. In 1896 R. Schöne found the text of this work in Istanbul, in a manuscript dating from the eleventh or twelfth century, and his son, H. Schöne, edited the work which was published, together with a German translation, in 1906.

The *Metrica* is concerned with *geodesy*, in the Greek sense of the term, that is in the sense of applied geometry. Unlike Euclid, Heron mainly deals with numerical examples, and calculates lengths, areas and volumes. The treatise comprises three books:

BOOK I—*Plane figures*. Introduction. (1) Rectangle. (2) Right-angled triangle. (3) Isosceles triangle. (4) Given the three sides of a triangle to decide whether a particular angle is acute, obtuse or a right-angle. (5) To calculate the height

---

* Bailly is mistaken in this assertion. The first book of the *Almagest* provided material for textbooks like the *Sphaera* of Sacrobosco. See p. 184. J.V.F.

of an acute-angled triangle in terms of the sides. (6) Calculation of the height of an obtuse-angled triangle. (7 and 8) Direct calculation of the area of a triangle in terms of the sides. This is our formula $A = \sqrt{s(s-a)(s-b)(s-c)}$, with proof. (9) Calculation of the area when the height is irrational. (10) A trapezium having one side perpendicular to its base. (11) Isosceles trapezium. (12) Acute-angled trapezium. (13) Obtuse-angled trapezium. (14 and 15) A quadrilateral containing a right-angle. (16) The same with a re-entrant angle. (17, 25) Regular polygons with 3, 5, 6, 7, 8, 9, 10, 11 and 12 sides. (26) Circles and circular rings. (27, 33) Segments of circles. As in the case of the regular polygons, Heron gives several approximate methods which belong to a continuous tradition going back to the Babylonians. (34) The ellipse. (35) The parabola. (36) The surface of a cylinder. (37) Of a cone. (38) Of a Sphere. (39) Of a segment of a sphere.

BOOK II—*Volumes.*—Introduction, General principles. Rectilinear solids. (1) Cone. (2) Oblique cylinder. (3) Oblique prism with a hexagonal base, called a *parallelepiped*. (4) Prism (semi-parallelepiped). (5) Pyramid. (6) Frustum of a pyramid. (7) Frustum of an oblique triangular pyramid. (8) Solid with two parallel but different rectangular ends ($\beta\omega\mu\iota\sigma\kappa\sigma\varsigma$, little altar). (9, 10) Frustum of a cone. (11) Sphere. (12) Segment of a sphere. (13) Spire (torus).* Heron attributes to Dionysodorus the theorem which gives the volume of the torus. (14, 15) Sections of cylinders. These are the results obtained by Archimedes in his letter to Eratosthenes: A cylinder of revolution is cut by a plane passing through a diameter of the base and tangential to the upper surface. This segment has a volume equal to one sixth of that of the prism of the same height with base formed by the square circumscribing the circular base of

---

* Heath uses 'tore'—the word means 'anchor ring'.

the cylinder. Two cylinders inscribed in a cube share a volume which is $\frac{2}{3}$ of the volume of the cube. (16–18) Regular polyhedra. (20) Indication of Archimedes's method of estimating the volume of irregular bodies (by means of the volume of water displaced).

BOOK III—*Division of plane figures.* This book is comparable with a similar work by Euclid, whose contents were summarized by Woepcke in 1851, working from an Arabic version. Introduction. (1) To divide a triangle in a given ratio by means of a line passing through one vertex. (2) By means of a line parallel to the base. (3) By means of a line through a given point on one side. (4) To inscribe in a given triangle another triangle, of given area, such that the three remaining triangles are equivalent to one another. (5) To divide a trapezium in a given ratio by means of a straight line through the meet of two sides. (6) By means of a straight line through a given point of the base. (7) By a line parallel to the base. It is perhaps interesting to note that when the Babylonians wanted to divide a trapezium with parallel sides a and b into two equivalent parts by a parallel line of length x they noted that $2x^2 = a^2 + b^2$. This fact alone suffices to show that there was a very ancient tradition of problems of dividing plane figures. (8) By a straight line passing through a given point of one side. (9) To divide a circle in a given ratio by means of another circle, concentric with the first. (10) To divide a triangle in a given ratio by means of a straight line passing through a given point on a produced part of the base. (11, 12) A quadrilateral by a straight line through a given point on one side. (13) The same problem but the point not on any of the sides. (14,15) Same problem for any polygon. (16) From a given point to draw a straight line which cuts off segments from two parallel lines, the segments to be measured from given points and to be such that their sum has some given value.

(17) To divide the surface of a sphere in a given ratio by means of a plane. This is Proposition 3 of Book II of Archimedes's *On the Sphere and Cylinder*. (18) To draw a straight line which cuts off one third of a circle. Heron notes that this problem 'is not rational' and that he can only give a good approximation. Euclid deals with a similar problem which can be solved exactly in many cases: To draw two parallel lines in a circle such that they cut off a given part of it. (19) To find a point such that when it is joined to the vertices of a triangle the triangle is divided into three equivalent triangles. (20) To divide a pyramid in a given ratio by means of a plane parallel to the base. It is here that Heron gives the remarkable method for the extraction of an approximate cube root which we shall discuss in Vol. 2, p. 150. (21) Same problem for a cone. (22) For a frustum of a cone. (23) For a sphere. This is Problem 4 of Book II of Archimedes's *On the Sphere and Cylinder*. It leads to a cubic equation, which many Greek and Arab mathematicians tried to solve.

2. TREATISE ON THE DIOPTRA. Heron's Dioptra ($\delta\iota o\pi\tau\rho\alpha$) is an instrument rather like the modern theodolite but without its optical apparatus, which is replaced by an alidade with open sights. The Greek text of Heron's work was first published by A.-J.-H. Vincent in 1858. Heron describes the apparatus and solves several problems involving levelling and alignment. As in the *Metrica* he derives the formula for the area of a triangle in terms of its sides.

3. THE PNEUMATICA.
4. ON THE ART OF CONSTRUCTING AUTOMATA.

These works concern Physics and Mechanics, as do the works on the construction of engines of war. Among Heron's works we should also mention the DEFINITIONS, which may be apocryphal, and a commentary on Euclid, of which only fragments have survived.

We now turn to another shadowy figure among the various Greek authors who made a personal contribution to mathematics or at least passed on an important tradition, namely Diophantus of Alexandria, who plays a very important part in the history of algebra.

We do not know for certain when he lived. He must be later than Hypsicles, whom he quotes, and earlier than Theon of Alexandria, who quotes him. He is at present thought to have been active about AD 250.

The Greek Anthology (*Anthologia Palatina*) contains a problem which relates to the length of his life:

> This tomb holds Diophantus. Ah, what a marvel! And the tomb tells scientifically the measure of his life. God vouchsafed that he should be a boy for the sixth part of his life; when a twelfth was added, his cheeks acquired a beard; He kindled for him the light of marriage after a seventh, and in the fifth year after his marriage He granted him a son. Alas! late-begotten and miserable child, when he had reached the measure of half his father's life, the chill grave took him. After consoling his grief by this science of numbers for four years, he reached the end of his life. [*Answer*: 84.]

We shall not concern ourselves here with Diophantus's most important work, the thirteen books on arithmetic, to which we shall have occasion to refer in our chapters on algebra. Diophantus's work was a very important force during the scientific Renaissance in Western Europe. It was in this work that men like Bombelli, Vieta and Fermat found what was perhaps the best of their inspiration.

With Pappus of Alexandria, who probably lived a little later than Diophantus, we enter what we may properly call the Age of the Commentators. Pappus is certainly later than Claudius Ptolemy, whom he several times quotes. He also wrote a commentary on the *Almagest*. A manuscript dating from the end of the tenth century, which is preserved in the library at Leyden, contains a marginal note, beside a passage relating to Diocletian, saying that

'Pappus wrote in his reign'. Finally, in a passage of his commentary on the *Almagest*, Pappus calculates 'the place and time of the conjunction which gave rise to an eclipse in Tybi in 1068 after Nabonassar', that is in October 320 (by the Julian calendar). Proclus, who died in Athens in 486, several times quotes Pappus in the course of his commentary on the First Book of Euclid. Consequently we can agree that it is reasonable to suppose Pappus lived in the first half of the fourth century AD.

His works tell us nothing about his life, except that he had a son named Hermodorus, and that he was a teacher. Suidas describes him as a philosopher: perhaps on the strength of works which have not come down to us, dealing with geography and the interpretation of dreams.

Pappus's most important work is the *Mathematical Collection*. A part of Book II and all the following books up to the eighth have come down to us. He is one of our best sources of information about Greek mathematics and the present work makes extensive use of his writings. Commandino's Latin translation of Pappus made a considerable contribution to the development of mathematics in the seventeenth century.

Theon of Alexandria, who observed a solar eclipse in his native city in 365, is famous for his commentary on Ptolemy's *Almagest* and for his edition of Euclid's *Elements*.

His daughter Hypatia, who is said to have written a commentary on Diophantus's *Arithmetic*, was one of the last pagan philosophers. She was killed by a Christian mob in a riot in March 415.

We know a great deal about another commentator, Proclus of Lycia, from a biography written by his pupil, Marinus of Neapolis. Proclus was born in Byzantium on 9 February 412, and died in Athens on 17 April 486. He was called Proclus of Lycia because both his father, Patricius, and his mother, Marcella, came from Xanthus, in Lycia.

He was taken to Xanthus in early childhood and received a good education there. He went to Alexandria, famous for its schools, and learned rhetoric from Leonas and grammar from Orion. He also attended the schools the Romans had opened in Alexandria, and studied jurisprudence there, as his father, an expert in the subject, had recommended him to do. Leonas was obliged to return to Constantinople, and took Proclus with him. On his return to Alexandria Proclus devoted himself to the study of philosophy, particularly that of Aristotle. His teacher was Olympiodorus the elder. He also began to study mathematics, attending the classes given by a certain Heron, not to be confused with the engineer we mentioned earlier. Finally, he went to Athens, where he was to remain for the rest of his life, and studied under Plutarchus son of Nestorius (Plutarchus's name is the same as that of the author of the *Lives of Famous Men*). Proclus became a pupil of Syrianus, who was a Neoplatonist, and of Asclepigena, the daughter of Plutarchus. He succeeded Syranus as director of the School of Athens, after the death of the Syrian philosopher Domninos. He thus earned the surname Diadochus, or 'Successor'.

Proclus is best known for his *Commentary on the Timaeus of Plato* and his *Commentary on the First Book of Euclid*. The latter book is (with the *Collection* of Pappus) one of our best sources of information about the development of Greek mathematics.

Marinus of Neapolis, a city in Palestine, was a pupil of Proclus. He was converted from Judaism to Greek Neoplatonism and succeeded his master as head of the school. He is best known for his preface to the *Data* of Euclid, which contains a learned discussion of the significance of the word 'datum' in the works of the various Greek mathematicians.

Eutocius of Askelon, a city in Palestine, was born in about 480. His Commentary on Archimedes's *On the Sphere and Cylinder*, *Measurement of a Circle* and *On Plane Equilibriums*, and another commentary on the first four books of the *Conics* of Apollonius, have all preserved valuable information about ancient mathematics.

5.3. The Hindus

Hindu mathematicians, who were in contact with their Babylonian, Greek and Arab counterparts, had a considerable influence on the development of our mathematical knowledge. So, although we shall not mention Chinese, Japanese or Mayan mathematics, which, however interesting in themselves, had no effect on the development of the study in Western Europe, we must pause a moment to consider the work of some of the scholars who lived beside the Ganges and the Indus.

The only very early work which need concern us is the *Sulvasutras* or cord-rules, which we shall discuss briefly in Chapter 4, Vol. 2. The three known editions of this work, by Apastamba, Baudhayama and Katyayana, are all written in verse.

They represent a state of knowledge in between what we attribute to the Babylonians and what we know of the mathematics of the Hellenistic period, and they seem to date from the fifth, fourth or third century BC.

Hindu mathematics probably exercised its greatest influence on the West between the sixth and the twelfth centuries AD. It is to this phase of Indian culture that we owe our decimal system of writing numbers, and hence modern techniques of calculation.

We shall consider only three individual mathematicians. The earliest, Aryabhata, was born, according to his own account, in the year 475 or 476 AD. He was thus contemporary with Eutocius, the Syrian commentator we mentioned above. He lived in Kusumaputra, now called Patna. His surviving work is called *Aryabhathiyam*, and is written in

Sanskrit verse. It is divided into four parts: *Celestial Harmonies*, *Elements of Calculation*, *Of time and its measurement*, and *Spheres*. In 1879 the Journal Asiatique published a French translation by Léon Rodet of the second part of this work.

Brahmagupta was born in 598. His works have been known in the West since 1817, thanks to an English translation made by H. T. Colebrooke. Michel Chasles gives a very detailed analysis of them in his 'Aperçu historique'.

Bhaskara was born in 1114, at Biddur in the Dekkan, and died about 1185. Like the two preceding scholars he was an astronomer. His works, like those of Brahmagupta, have been known in the West since 1817, again thanks to Colebrooke.

The work of these three Hindu mathematicians was particularly important for the development of Number Theory and of Indeterminate Analysis, the field in which the leading Greek scholar was Diophantus. The Hindu mathematicians' approach to the subject is, however, different from that of Diophantus.

It is also from the Hindus that we derive the use of sines instead of chords in trigonometry.

5.4. The Arabs

It was the Arabs who passed on to us most of the learning of Antiquity.

However, we must note an important general point: when we speak of 'the Arabs' we in fact mean all those scholars who wrote in the dominant language of their times and of their country. We do not intend to imply anything about their origin or their religion. Euclid, Archimedes, Apollonius, Diophantus and Ptolemy wrote in Greek. We cannot conclude from this that they were

Greeks. Before them others had written in Cuneiform, Sumerian or Akkadian, but this fact does not allow us to deduce anything about their race or their religious beliefs. From mediaeval times until the eighteenth century men of very different origins wrote in Latin. In the eighteenth century French was used in Berlin as well as in Paris; in the twentieth century English is used by scholars from India, Germany and Japan.

By 'Arab Science' we merely mean scientific works written in the Arabic language. Most of the mathematicians we mention were Muslims, and some were actually Arabs, but in general they were men of various races and were born in the countries round the Mediterranean, in places as far apart as Persia and Spain.

Al-khovarizmi, more properly called Abu Abd Allah Muhammad ben Musa al-Huwarizmi, was born in Huwarizmi, the modern Khiwa. This city lies to the south of the Aral Sea, to the east of the Caspian, and is now the capital of the Socialist Soviet Republic of Khorezm.

We know nothing definite about al-Khovarizmi's life. We know only that he worked in the library of al-Ma' mûn, who was Caliph from 813 to 833 (a little later than the reign of Charlemagne in the West). It is likely that he first worked on drawing up astronomical tables. His name, corrupted into algoritmus, algorisme, gave us the word algorithm, which has taken on a very wide meaning as designating any procedure of calculation.

There is no surviving copy of al-Khovarizmi's *Arithmetic*, the work which introduced the Hindu system of numerals to the Arabs and thence to Western Europe. All we have is an *Algoritmi de numero indorum* which seems to be a fairly faithful translation, and a *Liber algorismi de practica arithmetice*, which is an extended and more elaborate version of the same book. Both books are by Johannes Hispalensis, John of Seville.

It has, however, been established that positional numerals were introduced into the West in the twelfth century through the *Arithmetic* of al-Khovarizmi.

There is only one surviving manuscript of al-Khovarizmi's *Algebra*. Gerard of Cremona (Gherardo da Cremona) made a Latin translation of it, entitled *Liber Maumeti filii Moysi alchoarismi de algebra et almuchabala*. We shall give summaries of several chapters of this work in Chapter 3, Vol. 2.

The word *algebra* is derived from al-gabr, and al-gabr, al-hatt and al-muqabala are all technical terms with the following meanings: if one or both sides of an equation contain a negative term this is taken over to the other side so as to make all the terms positive, and the process is called algebra, *al-gabr*.

The operation of dividing both sides of an equation by the same number is called *al-hatt*.

Al-muqabala consists of combining similar terms on either side of an equation.

consider the equation $\quad 8x^2 - 4x + 6 = 6x^2 + 4$
by al-gabr $\qquad\qquad\quad 8x^2 + 6 = 6x^2 + 4x + 4$
by al-hatt $\qquad\qquad\quad 4x^2 + 3 = 3x^2 + 2x + 2$
by al-muqabala $\qquad\quad\ x^2 + 1 = 2.$

The question has thus been reduced to one of al-Khovarizmi's standard forms.

Tâbit ben Qurra is particularly famous as a translator. His complete name is Abû al-Hasan Tâbit ben Qurra ben Marwân al-Harrâni.* He probably lived from 827 to 901. He made a translation of Apollonius, and it is in fact only through this translation that Books V, VI and VII of the *Conics* have come down to us. He also translated Archimedes, Eutocius, and Theodosius, and made a thorough

* Harran, where he was born, is near Edessa. It is now a Turkish city on the frontier with Syria. Tabit was a Sabean.

revision of Ishâq ben Hunayn's translation of Euclid's *Elements*. This revised edition survives in two Latin translations, one anonymous and one by Gerard of Cremona, as well as in a twelfth-century translation by Moses Ibn Tibbon. Ibn Tibbon belonged to a family of Rabbis who came from Granada, but he himself lived in Lunel and Marseille. There were also Persian, Syrian and Armenian versions of Tibbon's work.

Abû al-Abbâs al-Fadl ben Hâtim al-Nayrîzî, famous among the Latins as Anaritius, died in 922. He wrote commentaries on Euclid and on Ptolemy.

Abû Alî al-Hasan Ibn al-Haytam, known as Alhazen in the West, was one of the greatest of the Arab mathematicians.

He was born about 965 at Bassora, in Iraq, lived mainly in Egypt, and died in Cairo about 1039. We know the titles of ninety-two of his works. He himself lists twenty-five:

'As for mathematical writings, my works number twenty-five: 1. A commentary on and summary of Euclid's *Elements of Geometry and Arithmetic*; 2. A collection of geometrical and arithmetical propositions taken from the treatises of Euclid and of Apollonius: in this work I classified and divided up the propositions and gave proofs derived from mathematics, computation and logic, so that my arrangement of the material does not follow that of Euclid and Apollonius; 3. A commentary on and summary of the *Almagest*, based on proofs; in this I have only made a small number of computations, but if God grants me time, and circumstances permit me to finish my work, I shall set about writing a very detailed commentary of this same work, and subject all my results to arithmetical computation; 4. A collection of the basic principles of computation, in which I have deduced all the basic principles of computation from those laid down by Euclid in his *Elements of Geometry and Arithmetic*; in my work I have shown how to solve problems of computation, using the double method of geometrical analysis followed by arithmetical checks, while at the same time I have avoided using the principles and technical terms proper to algebra; 5. A summary of Optics, taken from the two works of Euclid and Ptolemy; 6. A treatise on the analysis of geometrical problems; 7. A treatise on the analysis of arithmetical problems by algebraic methods, with proofs; 8. A complete treatise on the analysis of geometrical and arithmetical problems; though the part dealing with arithmetical problems does not include proofs it is founded on the principles of algebra; 9. A treatise on mensuration in the manner of the *Elements*; 10. A treatise on commerical computations; 11. An improved manner of building and excavating, in which I have explained all the techniques involved in terms of geometrical figures, even going so far as to introduce the three conic sections, the parabola, the hyperbola and the ellipse; 12. A summary of Apollonius's *Books on the Conic Sections*; 13. A memoir on Indian computation; 14. A memoir on the determination of the azimuth of Kiblah at any point of the inhabited world, with the tables I constructed, but without derivations of the procedures used; 15. A work on certain geometrical problems whose solution is required for the purposes of religion; 16. A letter addressed to several raïs, to encourage astronomical observations; 17. An introduction to Geometry; 18. A memoir on the refutation of the proof that the hyperbola and its two asymptotes approach one another indefinitely without actually meeting one another; 19. A reply to seven mathematical problems which were proposed to me in Baghdad; 20. A treatise on the use of analysis and synthesis by geometers, for the use of students, a collection of geometrical and arithmetical problems in my own arrangements and with my own solutions to them; 21. A treatise on the universal instrument, a summary of the treatise by Ibrâhim Ben Henân; 22. A memoir on the geometrical method of finding the distance between two points on the Earth; 23. A memoir on the basic principles of arithmetical problems and of the analysis of them; 24. A memoir to resolve a doubt about Euclid, with reference to the fifth book of his treatise on the *Elements of Mathematics*; 25. A memoir on the proof of the theorem Archimedes proposed on the trisection of an angle, a theorem he did not prove.' [1]

Alhazen is unduly modest in this summary of his work: for instance, his work in optics, did not merely consist of writing commentaries on Euclid and Ptolemy, but in fact brought about a revolution in the whole conception of the subject by replacing the Greek idea of visual rays emanating from the eye by the idea that luminous rays travelled towards the eye from the object. He also wrote a remarkable study of the way the eye works. A problem in geometrical optics is named after him: given a circular mirror and two points A and B find a point M on the mirror such that the ray AM is reflected along MB.

It is said that Alhazen, who was known to his contemporaries as a learned man, a philosopher and a physician, came to Egypt with the intention of straightening out the course of the Nile to make its periods of flooding more regular. The ruler of the country encouraged him, and put workmen at his disposal, but when Alhazen had toured the country and had admired the grandiose remains of its ancient civilization he gave up his project, having done little more than make a few attempts at straightening the river's banks in the region near Aswan. Then, in order to avoid disgrace, and possibly also punishment, he pretended to be mad until the particular ruler died.

Abú al-Wafâ Muhammad ben Muhammad ben Yahyâ ben Ismâîl ben al-Abbâs al-Buzgânî, or Aboul Wéfâ, was born in June 940 into a learned family in the province of Khorassan, in North-West Persia. He died in Baghdad in 997 or 998.

Aboul Wéfâ is most famous as an astronomer, but his mathematical work made a considerable contribution to the development of trigonometry. His commentaries on Euclid, Diophantus and al-Khovarizmi are now lost.

Abú Kamil Sugâ ben Aslam ben Muhammad Ibn Sugâ, al-hasib al-misrî (an Egyptian who made many calculations) must have lived before Aboul Wéfâ since he seems to have been active around 900. He developed al-Khovarizmi's algebra, and exerted a great influence on the work of Alkarchî. His work was later used by Leonardo of Pisa.*

Abú Rayhân Muhammad ben Ahmad al-Birûni (973–1048) was a general scholar who occupied himself with mathematics. He was born in Khiwa, like al-Khovarizmi, and died in Gazna in Afghanistan.

When al-Birûni was about twenty he went to Gurgân, to the Ziyarid court, where he wrote his *Chronology*. While he was there he met the great philosopher Ibn Sina (980–1037) who was called Avicenna in the West, where he exerted considerable influence on Mediaeval thought. The two men did not like each other.

---

* Sometimes known as Fibonacci. J.V.F.

About 1010 al-Birûni returned to his native city, to the court of Abû al-Abbâs al-Mamun II. However, the prince was assassinated, and in the unrest which followed Mahmud of Gazna seized control of the country. He took al-Birûni back to his capital as a prisoner.

Al-Birûni, a Shia Muslim, was converted to the orthodox Suni faith, of which Mahmud was a fierce supporter. We know very little about the time al-Birûni spent in India, where his master's military campaigns detained him from 1001 until 1024. On his return to Gazna, Mahmud's successors gave him a comfortable official position at Court, and he seems to have remained there for the rest of his life.

Al-Birûni's travels in India provided material for an important book entitled *Ta rih al-Hind*, in which he gives a full account of the geographical features of the country, and discusses the religious beliefs and intellectual achievements of the Hindus, quoting passages from the works of various authors, among them Aryabhata and Brahmagupta.

Al-Birûni also wrote an encyclopedia of Astronomy.

His geometrical work included an original method of deriving the formula for the area of the triangle $A = \sqrt{s(s - a)(s - b)(s - c)}$, a formula which, as we have seen, was used by Heron, though al-Birûni attributes it to Archimedes. He also derived the formula $A = \sqrt{(s - a)(s - b)(s - c)(s - d)}$ which gives the area of a quadrilateral inscribed in a circle, and he stressed the fact that the formula is only valid for this type of quadrilateral, a condition which is not mentioned by Brahmagupta and Bhaskara, who merely state the formula without proof.

Al-Birûni reduced the problem of inscribing a regular nine-sided polygon in a circle to that of solving the equation $x^3 = 1 + 3x$, and gave the solution in sexagesimal fractions: 1; 52, 45, 47, 13.

In decimal notation this is 1·8798352468, and the answer is correct to at least the first eight places.

Al-Karkhi (more correctly known as Abû-Bakr Muhammad ben al-Hasan al-hasib al Karhi) who died between 1019 and 1029, was important for his work in Algebra, which was for the most part based on that of Diophantus. He did not use positional notation in his arithmetical work and all numbers are written out in full.

Omar Khayyam (Abû al-Fath Umar ben Ibrâhîm al-Hayyâmi, Giyat al-dîn) who was born in Nishapur in Persia about 1040 and died about 1123 was famous both as a poet and as an algebraist. He also reformed the old Persian calendar. We shall quote three stanzas from his works, two rather fanciful and one more serious.

'O thou, slave of the four [elements] and of the seven [spheres of the heavens], thou canst hardly move for the constraints put upon thee by these four and these seven. Drink wine, for I have told thee a thousand times: thou hast no hope of returning here, once thou art gone thou art gone forever.

'My desire is to drink so much wine that its fumes will rise from the earth over my tomb, and the drinker who stirs my dust will fall down drunken with the smell of wine.

'The atoms of the Sphere which adorn the World, they come here, they leave and they return.

'On the skirts of the robe of heaven and in the innermost pocket of the earth,

'Their multitude is to be found: they will be reborn so long as God endures.'

Omar Khayyam's work on algebra was translated into French by F. Woepcke in 1851. In this very important work Omar Khayyam attempts to classify all equations up to the fourth degree according to the number of terms which they contain. Cubic equations are solved geometrically by finding the points of intersection of various conics.

We shall end this rapid survey of important Arab mathematicians—and we have left out many famous names—with a brief reference to Nasir al Din al-Tusi (Abû Gafar Muhammad ben Muhammad ben al-Hasan, Nâsir al-din, al-Tûsî, al muhaqqiq, that is, the investigator*).

He was born in 1201 and as a young boy he was carried off by the Hasîsîyûm (Assassins) and imprisoned in their fortress at Alamût, where there was a large library which he was allowed to use for his education. In 1256 the grand master of the Assassins surrendered to the Mongol leader Hûlâgû Khan, and al-Tusi found himself a prisoner of the Mongols. The heretical books in the library were burned, and the rest were later incorporated into the library of the observatory at Maraga, which al-Tusi had persuaded the Mongol ruler to found. In the event, al-Tusi's scientific skill, and particularly his skill in Astrology, enabled him to pursue a successful career in the service of his new masters. He was vizier to Hûlâgû Khan, the grandson of Genghis Khan, and Il-khan of Persia from 1256 to 1265. He was present at the capture of Baghdad by the Mongols in 1258.

Al-Tusi was director of the observatory at Maraga, in Iran, Azerbaidjan, south of Tebriz, from 1259 until his death in Baghdad in 1274. Two of his sons succeeded him in his post.

Some of al-Tusi's numerous works are written in Arabic but others are written in Persian.

5.5. The Byzantines

We have already described the development of Greek mathematics up to the time of Eutocius, who wrote commentaries on the works of Archimedes and Apollonius, and it is clear that no sharp distinction can be drawn between

---

* His surname, Nasir al Din, means defender of the faith. He was born at Tus, a town in Korassan.

Hellenistic and Byzantine mathematical work, since the latter followed on continuously from the former and was written in the same language.

The scholarly tradition of the Eastern Empire was on a far higher intellectual level than that to which learning was to sink in the West, but it was always far below the standards of the great Hellenistic period. In fact Byzantine work did not reach the level attained by Hindu mathematicians, and also fell short of the level attained by the Arabs.

We shall therefore content ourselves with mentioning a few important names.

Anthemius of Tralles, who lived a little before Eutocius, was one of the architects responsible for the church of S. Sophia in Constantinople. He died in 534. He is as famous as an architect as he is for having written a book about burning glasses. The book, in which Anthemius makes use of the properties of conics established by Apollonius, survives as a fragment.

After Anthemius's death, the work on S. Sophia was continued by Isidore of Miletus, whom some believe to have been the teacher of Eutocius.

Michael Psellus lived towards the end of the eleventh century, five centuries after the scholars we have just mentioned. He wrote on many subjects, and is supposed, perhaps incorrectly, to have been the author of a work, translated into Latin as *De quattuor mathematicis scientiis*, several editions of which were published in the sixteenth century. A letter written by Psellus contains some comments on the *Arithmetic* of Diophantus.

In the thirteenth century George Pachymerus (1242–1310) and Maximus Planudes wrote commentaries on Diophantus. Planudes, who was the Emperor Andronicus II's ambassador to Venice in 1297, probably lived from 1260 to 1310. He wrote a work about the Indian system of numerals, which proves that this system was known in the

Byzantine Empire by the thirteenth century, at about the same time as it was gaining acceptance in the West.

John Pediasimos, who was Keeper of the Seal of the Patriach of Constantinople in the reign of Andronicus III (1328–1341), wrote a textbook of Geometry rather similar to that of Heron.

Barlaam, who died in 1348, was a monk of the order of St. Basil, and a parish priest in Constantinople, though he later lived near Naples. He wrote a commentary on Book II of Euclid's *Elements*, and a *Logistics*, in which he explains how to make calculations involving integers and both ordinary and sexagesimal fractions.

Manuel Moschopoulos was a pupil and friend of Maximus Planudes. He described a method of constructing magic squares.*

Moschopoulos's work is dedicated to *Nicolas Artavasde Rhabdas*, who was alive in 1341, and wrote a work describing the Greek system of numerals.

---

* A magic square is a square array made up of the first n^2 integers arranged so that the sum of the numbers contained in any line or column or in either of the diagonals is always the same. Two examples are:

4	9	2
3	5	7
8	1	6

1	15	14	4
12	6	7	9
8	10	11	5
13	3	2	16

Fig. 5.1

In the seventeenth century there was a great vogue for constructing magic squares (see Plate VIII). Fermat showed extraordinary ingenuity at it.

Mansell

PLATE VIII. Albrecht Dürer's *Melancolía*. Note the magic square.

5.6. The West

The Western Empire broke up under pressure from invaders, and in losing its contact with Byzantium found itself cut off from the continuing intellectual tradition of Antiquity.

It retained only what had been passed on in Latin, which, as far as mathematics was concerned, meant only elementary techniques of computation and surveying.

The Latin scholars of late antiquity whose influence was to extend into the Early Middle Ages were Victor of Aquitaine, Cassiodorus, Boethius, Isidore of Seville and the Venerable Bede.

Victor of Aquitaine is famous for a *Canon paschalis* which he wrote in Rome in 457 : it shows that after a cycle of 532 years Easter day would once again fall on the same date. He also drew up a huge multiplication table, known as the Victorii Calculus, which we shall discuss in Chapter 1, Vol. 2.

Cassiodorus (Flavius Magnus Aurelius Cassiodorus Senator) was probably born in 490, in Brutium (the modern Calabria). He addressed a flattering oration to Theodoric, King of the Ostrogoths, and as a result was appointed quaestor in 507 and consul in 514. Around 540 he spent some time in Constantinople. He then retired to his native city of Scilacium (the modern Squillace) where he founded a monastery, at Vivarium. He lived to a great age, and was certainly still alive in 580 AD.

Cassiodorus's interests were wide : he wrote a *Historia Gothorum*, a work called *Institutiones divinarum et humanarum litterarum* (an encyclopedia of the seven liberal arts) as well as a *Variarum libri XII*. These works helped to pass on to the Middle Ages a small part of the arithmetic and geometry which had been known to the Ancient World. Cassiodorus's most important contribution, however, was in setting his monks to copy out manuscripts. Monasteries continued this practice for a long time, and it had considerable influence on the development of scientific and literary tradition.

Martianus Mineus Felix Capella, who was born in Madaura in Numidia had interests as wide as those of Cassiodurus. In 470 he composed a *Satyricon* or *De nuptiis Philologiae et Mercurii et de septem artibus liberalibus libri novem*, a work which deals in turn with grammar,

dialectic, rhetoric, geometry (including geography), arithmetic, astronomy and music (including poetry). It was
exceedingly influential during the Middle Ages, despite its
mediocrity.

Boethius (Anicius Manlius Severinus Boetius or Boethius)
was however, a figure much more representative of the last
throes of the Ancient tradition.

Boethius was born in Rome about 480 AD. His father
gained the favour of King Theodoric and was made consul
in 510, a dignity to which his two sons also attained in 522.
Boethius, who lost his parents at an early age, was brought
up by Symmachus, whose daughter he eventually married.
Boethius and Symmachus both held important offices
under Theodoric but fell into disgrace and were eventually
executed in Pavia in 524, after many years of imprisonment.

Boethius was a Christian, but he looked back nostal-
gically to the learning of Antiquity. He made Latin translations of several ancient works and compiled extracts from
others. The two books of his *De institutione arithmetica*
rework the material of Nicomachus's two books on
arithmetic, while the five books of the *De institutione
musicae*, which deal with music (and therefore in particular
with ratios and means) are compiled from the works of
Nicomachus, Euclid and Ptolemy. Boethius must also
have made a Latin translation or adaptation of Euclid's
Elements, but *Geometria Euclidis a Boetio in latinum lucidius
translata*, mentioned in several eleventh and twelfth century
manuscripts, is probably not by Boethius, since it seems to
date from the ninth or tenth century.

This apocryphal *Geometria* contains the definitions,
five of the postulates and three of the axioms from Book I
of the *Elements*, some definitions taken from Books II,
III and IV, the propositions from Book I (these are stated
without proof), ten propositions from Book II, some more
from Books III and IV, and finally a literal translation of

the proofs of Propositions 1, 2 and 3 from Book I, presented as the author's original work.

This somewhat slight work was reprinted in Paris in 1531, by Simon Collines, since it was then still used as a textbook in the colleges of the university.*

Saint Isidore, Isidorus Hispalensis, or Isidore of Seville, a very learned man, was born in Carthagena or Seville in about 560 AD and in about 600 became Bishop of Seville, where he died in 636. He wrote a treatise entitled *De natura rerum*, dedicated to Sisebutes, King of the Visigoths, and also encyclopedias of Cosmography, Astronomy and Meteorology, but he is best known for his *Etymologiarum sive originum libri XX*.

The Venerable Bede (673–735), a Benedictine monk born in Northumberland, where he spent almost all his life, was reputed to be one of the most learned men of his time and is frequently cited as an authority by later writers. He knew Greek, then a rare accomplishment in the West, and in his scientific work he turned for guidance to Pliny and Isidore of Seville. In a treatise called *De loquela per gestum digitorum* he explains a way of representing numbers up to 100 000 by using the hands.

Finger counting probably goes back to very remote antiquity, but no early references to the practice have survived. Apart from Bede, we owe our knowledge of the principles involved to Fibonacci (thirteenth century), Rhabdas (fourteenth century), and Luca Pacioli (fifteenth century). Many mediaeval miniatures and tapestries show us scholars, usually astronomers, counting on their fingers, which are bent in various ways, and this is how the fifteenth-century painter showed Boethius in the portrait we reproduce on p. 173. [The picture was painted for the Ducal Palace at Urbino, where it may still be seen.]

---

* Several editions of it were printed in Venice from 1497 to 1499, and in Basel in 1546 and 1570.

Alcuin was born in Yorkshire, probably in the year of Bede's death, 735, and he himself tells us of his education in York under the guidance of Egbert:

The learned Egbert gave drink to thirsty spirits from all the springs of knowledge. To some he taught the rules of grammar; for others he let flow the streams of rhetoric. Some he trained to debate at the Bar, and others to sing the songs of Aonia. He also taught them how to make music with the Castalian flute and to measure a lyric tread upon the heights of Parnassus. He also explained the harmony of the heavens, the chilling eclipses of the sun and moon, the five zones of the pole, the seven planets, the laws of heavenly bodies, their rising and setting, the violent movements of the sea, earthquakes, the nature of man, of cattle, of birds and wild beasts, and the different combinations of numbers and their various forms. He taught a sure way of calculating the date of the solemn return of Easter, and, above all, he drew the veil from the mysteries of Holy Scripture: he had gazed into the abyss of the Old Testament.

This passage provides a brief description of the education the monasteries gave to those destined for high office in the Church. Alcuin eventually succeeded his teacher as director of the school at York.

At Parma, in the course of a journey to Italy he made in 781 or 782, Alcuin was introduced to Charlemagne, and he was later to play an important part in the new Emperor's attempt to reorganize the educational system. He died at Tours on 19 May 804.

Alcuin's place in the history of Mathematics is as the supposed author of the *Propositiones ad acuendos juvenes*, one of the oldest examples of 'mathematical recreations'. The work is a collection of propositions, some arithmetical and some to do with surveying, and its attribution to Alcuin is not absolutely certain.

The author explains the importance of the number 6 in the creation of the world by the fact that 6 is a perfect number. The second rebirth of the human race is linked with the deficient number eight, since Noah's ark contained eight human beings, from whom the whole human race was to spring, which shows that this second origin of mankind was less perfect than the first, which had been based on the number 6.*

Gerbert (Pope Sylvester II) was born about 930 in Aurillac, where he also spent his youth. It seems that in about 967 he began the series of contacts with Arab scholars which were to prove so fruitful in later generations.

At this time the small Catalan town of Ripoll was the frontier post between the two worlds of Christendom and Islam.

Gerbert held the influential position of Scholiast (i.e. director of the diocesan school) of Rheims, from 972 to 982. He was later Abbot of Bobbio, in Italy, an abbey which possessed a very fine library. He returned to Rheims, became an adviser to Pope Gregory V in 996, and Bishop of Ravenna in 998. He was elected Pope on 2 April 999, and died on 12 May 1003.

It seems to have been Gerbert who introduced into the West the practice of making calculations by using marked discs (apices). This method, which had nearly all the operational advantages of positional arithmetic, was used in abacus calculations throughout the eleventh and twelfth centuries.

Gerbert and his contemporaries were much more adept at practical computation than they were at Geometry. For instance, Adelbold, Bishop of Utrecht, who died in 1027, wrote to his master Gerbert to ask him about a difficult problem: two different ways of calculating the area of a triangle had given two different results and Adelbold wanted to know which answer was the correct one. Gerbert's

---

* 6 is a *perfect* number (Euclid, *Elements*, Book IX). 8 is *deficient*: it is greater than the sum of its divisors

$$8 > 1 + 2 + 4$$

An *abundant* number (Nicomachus, Boethius) is less than the sum of its divisors. The first abundant number is 12:

$$12 < 1 + 2 + 3 + 4 + 6.$$

reply was to the point, but shows the elementary nature of the teaching of the time. An equilateral triangle with base 7 has height 6. (Gerbert recognizes that this height is only approximate.) Adelbold had used these values to calculate the area of the triangle 'geometrically', and had found it was 21. But if he calculated the area 'arithmetically', that is by constructing the seventh triangular number* he obtained an answer of 28.

In his reply Gerbert explains that the triangular number 28 represents a block of 28 squares, some of which protrude beyond the sides of the triangle, which explains the anomalous result.

This crude method of employing figurate numbers to calculate the areas of polygons can be traced back to the work of Roman surveyors. The tradition clearly either persisted until the eleventh century or had been rediscovered.

Paul Tannery drew attention to some letters, dating from about 1025, which were exchanged between Ragimbold, Grand Scholiast of Cologne, and Radolf of Liege, *magister specialis* and then *magister scholarum*.

The correspondence begins with a discussion of the meaning of the word 'interior' as used by Boethius in his commentary on the *Categories* of Aristotle: 'We know that the sum of the interior angles of a triangle is two right angles.'

It appears that Ragimbold had formerly discussed this problem with his teacher, Fulbert, at Chartres, and they had agreed that 'interior' must mean 'acute' and 'exterior' would mean 'obtuse'. Radolf, on the other hand, thought that 'interior' referred to angles drawn on a plane and 'exterior' to angles on the surface of a solid, such as a cube.

We shall not discuss the further course of the argument. What we have said already suffices to show that geometry

---

* See above, p. 42.

played no part in the education of Western scholars of the early eleventh century. It would, however, be rash to assume that the ninth-century master masons who built such fine Romanesque cathedrals as that of Pay-en-Velais were ignorant in matters of geometry, and contact with the Arabs, mainly through the Jewish communities of Spain and Provence, was in fact soon to lead to a slow revival of the intellectual tradition of Antiquity.

Rabbi Abraham Bar Hiyya ha-Nasi, a Jew born in Catalonia (perhaps in the town of Barcelona), lived in the later part of the eleventh century and the early part of the twelfth. His Hebrew title was Nasi (prince) but it is from the Arabic one, Sâhib al-Surta (captain of the guard), that he was given the name *Savasorda*, by which he is generally known. Savasorda, an accomplished linguist, learned from the works of Arab scholars but wrote all his own works in Hebrew, incidentally making a considerable contribution to the scientific vocabulary of that language. While living in Barcelona he worked together with Plato of Tivoli (see p. 172) who translated Arabic and Hebrew works into Latin.

Savasorda's scientific output includes writings on astronomy and a work on geometry, written in 1116, translated into Latin by Plato of Tivoli, and known as the *Liber embadorum*.

This work has much in common with the *Metrica* of Heron and the *Practica Geometriae* of Leonardo of Pisa, which we shall discuss below. It is divided into four chapters. The first contains the definitions, postulates and axioms from the first book of Euclid's *Elements*, and the arithmetical definitions from Book VII, as well as theorems connected with Greek algebraic geometry, together with numerical examples, and statements of elementary theorems on the equality of plane figures.

The second chapter, which deals with calculating areas, can be divided into five parts: squares and rectangles;

triangles; parallelograms, trapezia and other quadrilaterals; circles and parts of a circle; polygons and measurement on sloping ground. It describes al-Khovarizmi's method of solving quadratic equations, how to calculate the area of a triangle given the three sides, and gives a table of the chords of arcs of a circle.

The third chapter deals with the division of areas, somewhat in the manner of Heron's *Metrica*. The fourth considers the measurement of solids, and then gives instructions for carrying out practical work.

Abraham ben Meir Ibn Ezra, or Aben Ezra (in Arabic, Abû Ishâq Ibrahîm Ibn al-Mâgid), is often called Abenarus or Abraham Judaeus by the Latins. He lived from about 1090 until 1167. He was famous for his commentaries on the Bible and his work contains a strange mixture of rationality and mysticism. One of the works he translated from Arabic into Hebrew was al-Biruni's commentary on al-Khovarizmi's tables. The original text of these tables is now lost. In his preface, Aben Ezra throws an interesting light on the way Hindu numerals were introduced into the Arab world.

Aben Ezra left two original treatises on arithmetic, one of which concerns the properties of the numbers from 1 to 9. A work called *Liber augmenti et diminutionis vacatus numeratio divinationis ex eo quod sapientes Indi posuerunt, quem Abraham compilavit et secundum librum qui Indorum dictus est composuit* has been supposed to be a Latin translation of a work by Aben Ezra, but the attribution has also been questioned, and some scholars attribute the work to the Arab Abu Khamil. It deals with algebra, and makes systematic use of the methods of single and double false hypothesis (*hypothesis falsi*).*

The Jewish communities of Spain and Southern France played an important part in the revival of learning in the

---

* This will be explained in Vol. 2, Chapter 2.

West, but the work of the translators was of equal importance.

Ioannes Hispalensis or John of Seville or Avendeut (son of David) was a converted Jew who, it seems, made translations from Arabic into Castilian. His collaborator, Domingo Gundisalvo, then re-translated the works from Castilian into Latin. Sarton gives a list of translations: one work on arithmetic, thirteen on astronomy and astrology, one on medicine, seven on philosophy.

Robert of Chester lived in Spain and is mentioned as having been an archdeacon at Pamplona in 1143. His translation of al-Khovarizmi's *Algebra* is dated from Segovia in 1145.

Plato of Tivoli (Plato Tiburtinus) lived in Barcelona from 1134 until 1145. He translated several Arabic works on Astronomy as well as the *Liber Embadorum* of Savasorda.

However, the most prolific translator of the twelfth century was Gerard of Cremona (Gherardo da Cremona), who was born about 1114 and died in Toledo in 1187. He has left us eighty-seven translations of Arabic works. The ones concerned with mathematics include treatises by Autolycus, Euclid (the *Elements*), Archimedes, Apollonius, Hypsicles, Theodosius, Geminus, Ptolemy (the *Almagest*) and al-Khovarizmi.

Adelard of Bath probably had closer contacts with Arab civilization than did the other scholars we have mentioned. He was born in England, in about 1070, but spent much of his youth in Normandy and along the river Loire. One of his early works, the *Regulae abaci*, was certainly written before he travelled in the East. He left France between 1104 and 1107, on travels that took him to Salerno and to Sicily, where he stayed for some time. (He dedicated one of his works to the Bishop of Syracuse.) He is also known to have visited Antioch before returning to settle in France, in the city of Laon. He was still alive in 1146. He left a Latin translation of the fifteen books of Euclid's *Elements* which in the next century was to inspire the work of Campanus of Novara.

PLATE IX. Boethius, portrayed finger-counting by Justus van Ghent.

6. First Recovery

Me stato anchor referto da piu persone, che
un Lonardo Pisano, trasporto la pratica di
queste tre scientie, over Discipline Arith-
metica, Geometria, et Algebra, di Arabia in
*Italia.**
Nicolo Tartaglia, 1556.

Leonardo of Pisa was born about 1170 and is usually
known as Fibonacci. His name in fact seems to have been
Leonardo Bigollo—at least, that is what he is called in a
document concerning the purchase of an estate, including
a tower and farm buildings, in which he was acting as an
agent for his brother, Bonaccingo.

The surname Fibonacci, a contraction of 'son of
Bonaccio', indicates that one of Leonardo's forebears,
possibly his father, was called Bonaccio. The family had
been established in Pisa at least since the eleventh century.
The mathematician's father, who was employed as a clerk
or a notary to the Republic of Pisa, was eventually appointed
to serve at the customs post of Bougie, where his job must
have been somewhat like that of a modern consul, and in
about 1192 he summoned his son to join him, hoping to
familiarize him with the techniques employed in business
and commerce, particularly those of calculation.

---

* I have been told by several people that a certain Leonardo of Pisa
brought the practice of these three sciences or disciplines, arithmetic,
geometry and algebra, from Arabia to Italy.

This information is given by Leonardo himself, in the preface to his most important work, the *Liber Abaci*. After travelling in Egypt, Syria, Greece, Sicily and Provence he became convinced that the Indian system of calculation (modern positional arithmetic) was by far the best. He read Euclid, and when he came to write his own work he added material of his own to what he had learned from Arab sources. His *Liber Abaci* was finally published in 1202 and the second edition appeared in 1228.

This second edition is dedicated to Michael Scot,* Astrologer to the Emperor Frederick II. Leonardo's connection with the Sicilian court is interesting because Sicily played an important part in the transmission of Arab science to the West. The situation there was similar to that in Spain, and original Greek works sometimes found their way to the West from Sicily.

The Saracens had landed in Sicily in 827, and had controlled all the island after taking Syracuse, which they captured in 878. Between 1060 and 1092 the Normans gradually drove them out.

The Saracens, and after them the Normans, ruled with great tolerance. Both Christianity and Islam were practised freely and all three languages, Greek, Latin and Arabic, were spoken and understood.

Frederick II, the Hohenstaufen Holy Roman Emperor (1194–1250), was proclaimed king in 1198, under a Papal regent. His court was a very brilliant one, and he was an enlightened patron of letters and of science, as well as a man of some learning. His six-volume work on falconry, *De arte venandi cum avibus*, shows strong Arabic influence.

The *Liber Abaci* of Leonardo of Pisa contains fifteen chapters.

---

* *Michele Scotto ... che veramente*
Delle magiche frode seppe il gioco.
   (Dante, *Inferno*, XX, 116–17)
Michael Scot ... who indeed knew the game of magical trickery.

The first deals with positional numerals, and also with calculation on the fingers, as described by Bede.

The second chapter deals with the multiplication of integers, the third with their addition, the fourth with their subtraction, the fifth with their division. Each section contains worked numerical examples, using the numbers 9, 7 or 11. Leonardo also explains how to factorize numbers into primes, and gives the conditions for a number to be divisible by $2, 3, \ldots, 13$. He also deals with calculations involving money.

Chapters 6 and 7 are concerned with fractions. For example, there is a table showing how to express certain fractions in terms of fractions of the form $1/n$. This is reminiscent of Egyptian work.

Chapters 8 and 9 deal with commercial applications, Chapter 10 with problems of business and Chapter 11 with problems involving exchanges, and other first degree problems of indeterminate analysis.

Chapters 12 and 13 are concerned with problems involving the methods of single and double false hypothesis.

Chapter 14 deals with calculations which involve square and cube roots. It thus belongs to the tradition derived from Book X of Euclid's *Elements*.

The last chapter of the *Liber Abaci* is devoted to numerical problems in Geometry and to the solution of quadratic equations, using the methods employed by al-Khovarizmi.

Leonardo of Pisa's second work, *Practica Geometriae*, which appeared in 1223, is in the tradition of Heron's *Metrica* and the *Liber embadorum* of Savasordo. It is divided into eight chapters or sections, prefaced by an introduction containing definitions and a list of the units of measurement employed in Pisa. The first section of the work explains how to calculate the area of a square or a rectangle. The second describes how to construct a mean proportional, using a straight edge and compasses, and gives a proof of Pythagoras's Theorem, using similar triangles. The third section returns to the evaluation of areas, and deals mainly with triangles: it shows how to find the area of a triangle from the lengths of the three sides, beginning with a numerical demonstration using sides of 13, 11 and 20, and then deriving the formula in a completely rigorous manner. The proof is, however, not that used by Heron but is an Arabic one, the method used by the Banū Mūsā (i.e. the sons of Musa ibn Shaker). Leonardo also gives a method of calculating chords of a circle, and practical methods for computing areas which do not lie in a plane.

The fourth section deals with the division of figures, and belongs to the tradition of Euclid, Heron and Savasorda.

The fifth is concerned with extracting cube roots, and with the problems of doubling the cube, using the methods of Archytas, Plato and Heron. There follow rules for handling cube roots in multiplication, division, addition and subtraction.

The sixth section contains calculations of volumes, in particular that of a frustum of a pyramid.

The seventh part is entirely concerned with the use of the geometrical square,* a practical instrument which was used to determine distances and heights (see Plate X, p. 178).

The next sections present theoretical problems concerning regular pentagons and decagons, and rectangles inscribed in a given triangle. Finally, Leonardo describes how to obtain approximate rational solutions to the equation $x^2 + 5 = y^2$.

Another of Leonardo's works is entitled: *Flos Leonardi Bigolli Pisani super solutionibus quarumdam quaestionum ad numerum et ad geometriam pertinentium*, that is: The flower of solutions to certain problems concerning numbers and geometry. The author explains that he gave the work this

---

* Sometimes called a quadrat. H.G.F.

title because several of these problems, though thorny, are explained in a manner as delightful as flowers, and just as plants have their roots in the earth and grow until they burst into flower so these problems give rise to a multitude of others.

B.N.

PLATE X. Measuring a distance using a geometrical square or quadrat. From Oronce Finé, *Geometria Practica* (Strasbourg, 1543).

Of the fifteen problems considered, thirteen are linear, determinate or indeterminate. The two others were proposed to Leonardo by John of Palermo, a philosopher at the court of Frederick II. The first of them concerns the equation $x^2 + 5 = y^2$, which we shall discuss below, and the second the equation $x^3 + 2x^2 + 10x = 20.$* We shall come across

---

* See Vol. 2, p. 103.

John of Palermo again a little later on. Nothing is known about him; but the fact that he put several difficult problems to Leonardo in the course of a kind of tourney, which was sometimes held in the presence of the Emperor, proves that there were cultivated mathematicians living in Italy in the early thirteenth century.

Leonardo has also left us an *Epistola Leonardi ad magistrum Theodorum Phylosophum domini Imperatoris*, which contains linear problems of indeterminate analysis and a rather curious geometrical problem (Fig. 6.1).

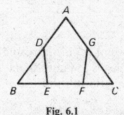

Fig. 6.1

In an isoceles triangle *ABC*, such that *AB = AC = 10* and *BC = 12*, it is required to construct a pentagon *ADEFGA*, with sides of equal length. Taking as his unknown length *AD = x* Leonardo obtains the equation:

$$x^2 + (36 + \tfrac{4}{7})x = 182 + \tfrac{6}{7}.$$

(His notation is naturally not what we have used here, and the equation is written out in full.) He thus considers that the problem has been reduced to an algebraic rule, and since the equation does not have a rational solution he gives an approximate solution, in sexagesimal fractions: $x = 4$; 27, 24, 40, 50.

The *Liber Quadratorum*, or *Book of Square Numbers*, is the fifth of Leonardo's known works, and it dates from 1225. The preface reads:

When, My Lord Frederick, most glorious prince, Master Dominic brought me to Pisa, to your Excellency's feet, Master John of Palermo made my acquaintance and asked me whether it was not as much a matter of geometry as of arithmetic to find a square number which if increased or decreased by five would still give a square number. Having thought about the solution to this problem, which I had already worked out, I satisfied myself that this solution depended on many properties of square numbers and their relationships. Having, moreover, heard from those in Pisa, and from others who returned there from the Imperial Court, that Your Majesty had deigned to read my book on numbers, and that Your Majesty sometimes took an interest in subtle points of geometry, I remembered the question that was put to me at Your Court by Your Philospher, and taking it as my theme, I wrote the present work, which I have called the *Book of Square Numbers*. I thus beg your indulgence if my work contains anything not altogether accurate or to the point; for it is an attribute rather of God than of man to be able to remember everything and to make no mistakes, and no-one can be completely free from error or always on his guard.

The most important problem treated in this short work (twenty propositions) is to find rational solutions to the set of simultaneous equations:

$$x^2 + u = y^2$$
$$x^2 - u = z^2.$$

The problem, which is not an easy one, is like those treated by Diophantus.

A marble inscription has survived recording the decree passed by the city of Pisa in 1240 fixing the remuneration to be paid to government accountants. Until then Leonardo had not been paid for his work. This inscription is our last document on the first original mathematician of the Western world.

The society in which Leonardo of Pisa lived was that of free cities almost entirely devoted to commerce. There were to be many such communities in Italy, in France, where Lyon was one of the most typical, in Flanders and in Germany, and they were always to play an important part in the teaching of practical mathematics. They even often appointed officials (called *Rechenmeister* in German-speaking communities) to supervise the teaching of mathematics, and for centuries the methods of teaching were similar to those employed by Leonardo.

We have, however, seen that around 1225 Leonardo was also in contact with another section of society, the cultured circle of the 'philosophers of the Emperor's court', which had links with the universities.

The history of the universities is well known: they had come into being in the course of the preceding century—the university of Bologna near the beginning of it, and the university of Paris around the middle years—but their charters date only from about 1200, during Fibonacci's own lifetime. Paris received its charter in 1200, Oxford and Cambridge in 1214 and 1231 respectively, Padau in 1222 and the University of Naples, the first State University, was founded by Frederick II in 1224.

The universities paid very little attention to mathematics, but the thirteenth-century mathematicians we are about to discuss did in fact all gravitate towards universities.

The work of Jordanus Nemorarius was comparable with that of Leonardo, but little is known of his life. The eminent historian of science George Sarton believes that Nemorarius was born in Westphalia in the second half of the twelfth century, was teaching in Paris in 1220, became a general of the Dominican order in 1222 and died in 1237, on board ship while returning from a pilgrimage to the Holy Land.

We shall not be concerned here with Jordanus's very important contribution to the development of mechanics.

In his treatise on the *Planisphere* Jordanus makes a stereographic projection of the sphere on to a plane tangent to the North pole, the eye being placed at the South pole. In Ptolemy's treatise of the same name, the projection is made from the South pole on to the plane of the equator.

Synesius of Cyrene, born about 370, who was a pupil of Hypatia of Alexandria, attributed the invention of this procedure to Hipparchus.

Ptolemy's work survives only in an Arabic version by Abû al-Qâsim Maslama ben Ahmad al-Magrîtî of Madrid, who died in Cordoba around 1007. This version was translated into Latin in the second half of the twelfth century, either by Rudolphus of Bruges or by his teacher Hermannus Secundus.

Ptolemy only proves that the stereographic* projection of a circle is a circle in particular cases, in fact he only proves it for cases he intends to use. Jordanus, however, gives a proof for the general case.

The treatises by Ptolemy and Jordanus were published together in editions printed in 1507, 1536 and 1538.

Curtze's edition of *Jordani Nemorarii Geometria vel de Triangulis lib IV* was published in 1887.

The first book of *De Triangulis* contains definitions, including a general discussion of continuity, and thirteen elementary propositions on the plane triangle, all based on Euclid.

The second book contains nineteen propositions, several of which are concerned with the division of areas by straight lines, in the manner of Heron, Savasorda and Leonardo of Pisa, but Jordanus does not merely copy his predecessors: he introduces new and rigorous methods of solving some of the problems.

The third book contains twelve propositions on areas and chords of circles. The fourth, which contains twenty-eight statements of propositions, considers regular polygons inscribed in a circle or circumscribed about it, as well as

---

* Chasles notes that the term 'stereographic projection' was first used by Aguilon, in his treatise on Optics, *Aguiloni Opticorum libri sex* (Paris, 1613).

the problem of doubling the cube and that of trisecting the angle. It also includes an approximate formula for finding the square of the side of a regular inscribed polygon, which we should write in modern notation as:

$$c^2 = \frac{36}{n(n-1)+6} r^2.$$

In this formula c is the side of the polygon, r the radius of the circle, and n the number of sides.

The tradition of using such approximate formulae goes back to Babylonian times and continues to the present day.

Curtze's commentary on Jordanus's *Tractatus de numeris datis*, published in 1891, drew attention to this important work.

The *De Numeris Datis* is a treatise on algebra inspired by the *Data* of Euclid. It is divided into four books and contains one hundred and fifteen problems, some abstract and some numerical, and all either linear or quadratic. The numbers are written in Roman numerals.

For example, the third proposition of Book I states that:

If a given number is divided into two parts such that the product of these parts is given, then each of these parts is given.

Let the given number abc be divided into ab and c and let d be the given product of ab by c. Similarly let the product of abc by itself be e. Let us take four times d, which we shall call f, and when we subtract it from e we are left with g. This number, g, will be the square of the difference between ab and c. Let us take the root of g and call it h, say. Then h will be the difference between ab and c. Since h is given, c and ab are also given.

This operation can easily be shown as follows. For example, let X be divided into two numbers such that their product is XXI. Four times this number is LXXXIIII. This must be subtracted from the square of X, that is from C, and the remainder is XVI. The square root of this is four, and this is the difference. We take this from X, and the remainder, which is VI, is then divided by two. The result is III, which is the small part, and the large one is VII.

Jordanus's *Arithmetica decem libris demonstrata* printed in 1496 and 1503 by Jacques Lefèvre of Etaples, deals with theoretical arithmetic, and has much in common with Euclid's arithmetical books, and with the works of Nicomachus and Boethius. For instance, Jordanus notes: the product of two consecutive numbers is neither a square nor a cube. Any multiple of a perfect or abundant number is abundant. Any factor of a perfect number is deficient. All abundant numbers are even (which is untrue).*

We know almost nothing about John Sacrobosco, also called Johannes de Sacrobosco, John Holywood and John of Halifax. Paul Tannery [1] writes: 'As an example of how little we know of some important matters I shall describe the case of Johannes de Sacrobosco, who lived in the mid thirteenth century. His works were highly esteemed by the students of the Arts† and about a hundred years after his death the University erected a tomb for him in the cloisters of the convent of the Mathurins.

'In fact, we know almost nothing about him. There is not even any proof that he really taught in Paris, since his name does not appear in any of the University documents referring to teachers of the Arts. I am often tempted to imagine him as a monk who wrote his *Sphaera* in his cell and taught, at most, only a few novices. His name might therefore only be preserved in some forgotten records of deaths, without even a mention of which year he died.'

The first of John Sacrobosco's works that we shall mention is his *Tractatus de arte numerandi*.

---

* For the definitions of the terms used see above, pp. 101, 168.

† The students and teachers of the Faculty of Arts, where after studying the Humanities scholars then devoted three or four years to the study of philosophy.

The work is in verse,* and is a sort of compendium of rules for carrying out calculations using Arabic numerals. There are no proofs, and no numerical examples. The various operations described are: Numeratio, Additio, Substractio, Mediatio (i.e. division by two), Duplicatio, Multiplicatio, Divisio, Progressio Extractio. The section called Progressio deals only with the summation of integers, and of pairs of successive even or odd numbers. Extractio deals with the taking of square and cube roots. The various rules that are given apply only to integers.

This sort of work was designed for oral instruction: the pupil would need to learn to recite it by heart, and the teacher would comment on it, giving explanations, if not actual proofs, as well as numerical examples. Among other commentaries on this work there is one by Petrus of Dacia, which was written in July 1291.

Sacrobosco's *Treatise on the Sphere* is the equivalent of a modern set of revision notes on cosmology. The introduction reads:

> The treatise on the sphere we divide into four chapters, telling, first, what a sphere is, what its centre is, what the axis of a sphere is, what the pole of the world is, how many spheres there are, and what the shape of the world is. In the second we give information concerning the circles of which this material sphere is composed and that supercelestial one, of which this is the image, is understood to be composed. In the third we talk about the rising and setting of the signs, and the diversity of days and nights which happens to those inhabiting diverse localities, and the division into climes. In the fourth the matter concerns the circles and motions of the planets, and the causes of eclipses.

This very elementary work had a lasting success. It went through many editions both in manuscript and in print, and commentaries were written on it as early as the thirteenth century; for example, there is one by Robert Anglès, a lecturer in Arts at Montpellier. Another was written at the

---

* The first two lines read:
Haec Algorismus, ars praesens dicitur, in quà
Talibus Indorum fruimur bis quinque figuris.

beginning of the sixteenth century by Jacques Lefèvre of Etaples, and another at the end of the same century by Clavius.*

Sacrobosco's works, which include, for example, a treatise on the Golden Rule and another on the quadrant,† may not be as scientifically significant as Leonardo's or Jordanus's, but they are based on sound scholarship and are perfectly adapted as aids to the teaching of elementary mathematics.

Vincent of Beauvais, a Dominican, who was born about 1190 and died a little after 1260, was a tutor to the children of King Louis IX of France (St. Louis). He wrote a huge encyclopedia, called the *Speculum Mundi*, or *Speculum Maius*, one edition of which, printed in Strasbourg in 1473, consists of ten folio volumes. It contains passages from Euclid, Aristotle, Vitruvius (the great Roman architect), Boethius, Cassiodorus, Isidore of Seville, al-Fârâbi and Avicenna. It also contains, under the heading 'Algorismus', a very clear explanation of the modern system of numerals, including the zero. Geometry, however, is reduced to definitions and a few elementary ideas.‡

---

* Maurolico did not, however, have a high opinion of this work: 'Quamquam hodie sphera Ioannis de Sacrobosco nullo non in gymnasio, quasi consummatissimum opus summa est auctoritas: Nec mirandum est, si rudis homo inter grammaticos aut dialecticos in astronomia peritus existimatur: nam luscus inter caecos oculatissimus est.' (We need not be surprised that the *Sphaera* of John Sacrobosco has considerable authority in all the schools, and that this ill-educated man passes among Grammarians and Logicians as one well-versed in Astronomy: in the kingdom of the blind the one-eyed man is king.)

† The Golden Rule is the set of rules which fix the dates of Church festivals. The quadrant is a surveying instrument which enables the user to work out the distances to inaccessible points. It can also be used to find altitudes (see also p. 177).

‡ One passage of the *Speculum* foreshadows a famous idea of Pascal's: 'Empedocles quoque sic Deum diffinire fertur: Deus est sphaera cujus centrum ubique, circumferentia nusquam.' (It is also reported that Empedocles defined God as follows: God is a sphere with its centre everywhere and its circumference nowhere.)

William de Moerbeke, a Fleming, is mainly famous for his translations of the works of Greek mathematicians. He knew Thomas Aquinas, who persuaded him to make a translation of Aristotle's *Physics*. Roger Bacon did not like Moerbeke and referred to him as 'William Fleming who has a high reputation at the moment, though all the educated men of Paris realize he knows nothing of the sciences in their original Greek, for all that he prides himself on having such knowledge.' Moerbeke had been to Greece several times, and had been appointed Archbishop of Corinth in 1278, though the appointment did not mean that he had to live in the city. His translation of Archimedes's *On Floating Bodies*, the work which contains the famous statement of Archimedes's Principle, was made in 1296 and entitled *De iis quae in humido vehuntur*. It was the only known version of Archimedes's work until 1906, when Heiberg found in a library in Istanbul the Greek manuscript which Moerbeke had used. The manuscript had been in the possession of Leo of Thessalonica (ninth century) and had found its way to the Sicilian court, then to Rome where Moerbeke had used it, and then to Venice, into the hands of the humanist Giorgio Valla (1430–1499). In 1543 Niccolò Tartaglia, whom we shall discuss later, published William de Moerbeke's translation in Venice, under his own name. Such practices were common at the time.

Witelo, better known as Vitellion, wrote a treatise on Optics which remained famous up to the time of Kepler. The latter, indeed, wrote a work on the same subject called *Ad Vitellionem Paralipomena*.* Vitellion's *Perspectiva*, which is dedicated to William de Moerbeke, can be dated to 1270. Like similar works by Roger Bacon, it is in the tradition of

---

* *Ad Vitellionem Paralipomena, quibus Astronomiae Pars Optica Traditur*, usually known as *Astronomiae Pars Optica*. The title means 'Things are left out by Witelo which relate to the optical part of Astronomy.' The work was published in 1604. J.V.F.

the work of Alhazen. Vitellion's grasp of mathematics seems to have been better than Bacon's.

In his dedication to William de Moerbeke, Vitellion refers to himself as 'filius Thuringorum et Polonorum', so he was of both German and Polish descent, and perhaps came from a Polish city which contained a prosperous German community.

Campano of Novara (Campanus) was a contemporary of the two preceding scholars. He was chaplain to Urban IV, who was Pope from 1261 to 1281. Campanus is famous for his Latin edition of Euclid's *Elements*, which was partly based on a translation by Adelard of Bath, who had made extensive use of Arabic sources. Campanus's work was printed in Venice in 1482.

We must also mention Yakob ben Mahir Ibn Tibbon, generally known as Don Profiat or Prophatius (Profacius) Judaeus. He was born in Marseilles in about 1236 and belonged to a famous family of Rabbis. He lived mainly in Montpellier, where he died in about 1305. He made Hebrew translations of the *Sphere in Motion* of Autolycus, the *Elements* and the *Data* of Euclid, the *Sphaerica* of Menelaus and various Arabic works. He invented a quadrant known as the quadrans judaïcus or quadrans novus,* and drew up an Almanac, which was very soon translated into Latin and was praised by Copernicus, Clavius and Kepler.

We shall only mention one fourteenth century scholar: Nicole Oresme.

Oresme was born in Normandy, near Caen, around 1323 and is mentioned as being in Paris, at the Collège de Navarre† from 1348 onwards, first as a student, then as a

---

* See p. 185, note †.

† The Collège de Navarre occupied the site of the modern École Polytechnique. It was founded in 1304 by Jeanne de Navarre, the wife of Philippe le Bel. It became a very aristocratic college (Henri III, Henri IV and the Duc de Guise all attended it at the same time) and Richelieu and Bossuet were among its later pupils. The King of France was the first benefactor of the College and the money from his benefaction was given to enable the College to buy birches for beating the pupils.

professor, and as Grand Master from 1356. He left the college in 1361, on being appointed Dean of the Chapter of Rouen. He became Bishop of Lisieux in 1377, and died on 11 July 1382.

He had close connections with the Court, and King Charles V did him the honour of attending the ceremony at which he was enthroned as Bishop.

Nicole Oresme's scientific work is written partly in Latin and partly in French, and he therefore played an important part in the history of the French language.

Manuscript 565 of the French collection of the Bibliothèque Nationale in Paris is richly decorated with magnificent miniatures and ornamental capital letters. As its first item it contains the charming *Treatise on the Sphere* (*De la Sphère*), which begins as follows:

To know the shape of the world and the disposition of its parts, the order and number of the elements and the movement of the heavenly bodies is an essential part of the education of any man of free condition and noble mind. And it is a fine thing and delightful and profitable and honest and also necessary to know more, and more particularly for the purpose of astrology. So that the human mind could understand such things more easily, learned men of ancient times invented among other things an instrument known as the material or artificial sphere, which one can look at from all sides and can turn round and use to display all the details of the motion of the world, and the sky as well. Like such an instrument, I want to tell everyone fully and in the common tongue what all men ought to know, without going into the scholarly proofs and subtleties that astrologers use, so I shall divide this work into chapters.

The work contains a little problem reminiscent of Jules Verne, despite the fact that it was written during the Hundred Years War. Robert, Peter and John are all in Paris, and Robert remains there. Peter sets off to go round the world, travelling always to the East, and John sets off on the same journey, but travelling always to the West. They meet again in Paris, where Robert says that their day of meeting is Sunday, while Peter maintains it is Monday and John that it is Saturday.

After the *Treatise on the Sphere*, a popular work intended for laymen, the manuscript next contains Oresme's important commentary on an apocryphal work of Aristotle, the *Treatise on the Sky and the World* (*Traité du Ciel et du Monde*).

We shall quote two passages from this work. The first concerns the nature of infinity:

> And thus outside the sky there is a space that is empty and incorporeal, in a way other than that in which a space may be filled and corporeal, just as that duration called eternity is different from temporal duration, even if that duration were to be perpetual. ...
>
> And that space I have referred to is infinite and indivisible and is God's immensity, and is God Himself, just as God's duration, called eternity, is infinite and indivisible and God Himself.

In the second passage, having declared that 'one cannot prove by any experiment that the Sky moves round in a day and not the Earth' Oresme adds:

> If a man were in the Sky, supposing it moves round in a day, and if that man, carried round with the Sky, could clearly see the Earth with its mountains, valleys, rivers, cities and castles, it would seem to him that the Earth moved round in a day, just as the Sky seems to move to us who are on the Earth. And similarly, if the Earth moved with this daily motion and not the Sky, it would seem to us that the Earth was at rest and the Sky in motion.

Oresme's other French works include translations of Aristotle [*Ethics*, *Politics*, *Economics**], a translation of the Bible, a translation of Petrarch's *De Remediis Utriusque Fortunae* and an important *Treatise on the first invention of money*, a work he first wrote in Latin and then translated into French at the request of Charles V.

Fourteenth century French was still a language of the people, concrete and very poor in abstract terms, so translators like Oresme were compelled to invent new words. In the *Treatise on the Sphere* Oresme in fact takes his readers into his confidence about his philological difficulties, which he resolves by transcribing words from Latin—the learned language in which he thought, as did all the scholars

---

* The King was interested in these works. In one of his inventories he says that the *Politics* and the *Economics* are 'very necessary' to him 'and for good reason'.

of his time—and then adapting these abstract terms to look like French. He apologizes for doing this, and even gives a glossary of the new terms at the end of his work. Oresme's various works introduced nearly three hundred words into the French language.

Before we turn to Oresme's mathematical works let us note that as a devout Catholic he was firmly opposed to astrology, a pseudo-science which still finds credence among some of our own contemporaries. In the twentieth century astrology no longer has any official support and is only practised by a few honest visionaries and a large number of charlatans, but in the fourteenth century it was a recognized science, and was, for instance, taught at the University of Paris, as the following anecdote shows:

On 12 February 1358 the Faculty of Arts met in the Church of Saint-Julien-le-Pauvre, which is still extant, on the left bank, very close to Notre-Dame. Albert of Saxony, a scholar of the same stature as Oresme and one who made considerable contributions to mechanics, asked permission to read Aristotle's *Politics* to his pupils on saints days, when there should not, in principle, be any lectures. Robertus Normannus (who is now completely forgotten) asked permission to read on the same days two works on astrology: the *Quadripartitum* of Ptolemy, and the *Centiloquium*, then thought to be by Ptolemy but in fact no more than a summary of the first work. Permission was granted to both teachers.

Oresme's treatise on *Latitudinibus formarum*, written in 1361, was printed in 1482, 1486, 1505 and 1515, and was used as a textbook in the universities of Cologne, Vienna and Ingolstadt. The ideas it expounds are not put forward as the author's own, and indeed they seem rather to have arisen in the scientific circles of Oxford.

Oresme represents properties graphically, using two dimensions: longitude and latitude. For example, extent in time is represented vertically, as longitude. The intensity of the property is represented as latitude, perpendicularly to its extent. If the property represented is velocity, the property which causes a body to travel a greater or less distance in a given time,* and if that velocity is uniform, the graph will be a straight line parallel to the axis of longitude, and *the distance travelled will be represented by a rectangle*.

If the velocity is 'uniformly non-uniform' the graph will be a straight line at an angle to the axis of longitude, and the distance travelled will be shown as a trapezium. The area of this trapezium will be the same as that of a rectangle with the same base and height equal to the mean height of the trapezium, from which we can deduce the very important fact that the distance travelled is the same as that which would be travelled by a body which had a uniform velocity equal to the arithmetic mean of the two extreme velocities.

Oresme's work was accepted universally until the seventeenth century, when it stimulated Galileo, Beeckman and Descartes to carry out further investigations into the properties of bodies in free fall.

The graphs drawn by Oresme and his contemporaries were naturally rather crude, more qualitative than quantitative, being made up of segments of straight lines or arcs of circles. Oresme noted the *analytical* point, which Ptolemy had already remarked upon in Astronomy, that any property only varies very slowly close to a maximum or minimum. Kepler later made the same observation. Fermat was to use algebraic methods as an *analytical tool* for finding maxima and minima. We shall return to this point later.

---

* It is perhaps of interest to note that in none of the work of seventeenth century applied mathematicians, except perhaps in that of Wallis, do we find the modern definition of velocity as the quotient of distance travelled divided by time, or the limit of this quotient as the time interval tends to zero. Even for Euler, in *Letters to a German Princess*, 'velocity is that well-known property whereby one says that in a certain time a body travels a greater or less distance in space'.

Another mathematical work by Oresme, the *Algorismus proportionum* was first printed in 1863, as a result of the work of Curtze. It uses a very interesting system of notation. For example, Oresme explains, in simple Latin, that:

'A half should be written $\boxed{\frac{1}{2}}$, a third $\boxed{\frac{1}{3}}$ and two thirds $\boxed{\frac{2}{3}}$, and so on.

The number above the line is called the numerator and the one below it the denominator. Double proportion will be written as 2^{1a}, triple as 3^{1a} and

so on. The sesquialter ratio is written $\boxed{\dfrac{p \quad 1}{1 \cdot 2}}$, sesquiquartus $\boxed{\dfrac{p \quad 1}{1 \cdot 4}}$. Super-

bipartiens tertius is written $\boxed{\dfrac{p \quad 2}{1 \cdot 3}}$, and duplus superbipartiens quartus

$\boxed{\dfrac{p \quad 2}{2 \cdot 4}}$, and so on. Half of double (duplus) proportion is written $\boxed{\dfrac{1 \quad p}{2 \cdot 2}}$,

a quarter of the duplus sesquialterus $\boxed{\dfrac{1 \cdot p \cdot 1}{4 \cdot 2 \cdot 2}}$ and so on'.

The last part of this passage is particularly interesting. We shall attempt to express it in modern notation.

If the 'proportion' of a quantity b to another quantity a is what we call 'double' (duplus) then we write $b = a \times 2$.

Oresme merely writes our multiplier as $\boxed{\dfrac{p}{2}}$.

What he writes as $\boxed{\dfrac{p \quad 1}{1 \cdot 2}}$ we should write $b = a \times [1 + \frac{1}{2}]$,

and his $\boxed{\dfrac{p \quad 2}{2 \cdot 4}}$ would be written $b = a \times 2[2 + \frac{2}{4}]$.

But when he divides a proportion by two, and writes $\boxed{\dfrac{1 \quad p}{2 \cdot 2}}$ we must understand this as $b = a\sqrt{2}$ or $b = a2^{1/2}$,

and when he writes $\boxed{\dfrac{1 \cdot p \cdot 1}{4 \cdot 2 \cdot 2}}$ we must read it as

$$b = a\sqrt[4]{2 + \tfrac{1}{2}} \quad \text{or} \quad b = a(2 + \tfrac{1}{2})^{1/4}.$$

This subject also interested the Jewish astronomer and mathematician Immanuel ben Jacob Bonfilius (Emmanuel Bonfils) who lived in Tarascon from 1340 until 1377. In Bibliothèque Nationale in Paris there is a Hebrew manuscript by Bonfils in which the author discusses the theory of exponents, exactly the subject of Oresme's work. Bonfils even introduces decimal numbers, which were not to come into common use until the end of the sixteenth century, after Stevin had carried out his completely independent work on them.

The fifteenth century saw the appearance of several vernacular treatises on arithmetic, designed for the use of tradesmen. Some of them were of very high academic standard.

In France one such treatise was written by Nicolas Chuquet, but it remained in manuscript form until it was discovered by Aristide Marre in the Bibliothèque Nationale in Paris in 1880. The part published by Marre is called the *Triparty en la Science des Nombres* and was written in Lyon in 1484 (see Plate I).

It begins:

This book, written in honour of the Glorious and Holy Trinity, is divided into three parts. The first treats of numbers as they may be enumerated, added, subtracted, multiplied and divided. And also of their proportions, progressions and other properties. The second part treats of the roots of numbers and the third is a book of primes or of the Rule of primes. The first part contains several chapters which appear in order as we describe below. The first of them is:
Enumeration.

To enumerate is to express the meaning of a number by representing it in ordinary figures or in words. To be able to enumerate things we need to know the ten figures used for this purpose, which enable us to write down any number. They are 0, 9, 8, 7, 6, 5, 4, 3, 2, 1. The first on the right denotes or signifies one. The next to the left denotes two, the third three, the next four, and so on.

The most interesting part of the *Triparty* is its 'third part', the 'book of primes': a treatise on algebra which employs an exponential notation very much ahead of its time. We shall return to this later. The last folio of the manuscript, numbered 147, ends with the words:

And also since it was written by Nicolas Chuquet, a Parisian, doctor of Medicine, I name it the 'triparty de Nicolas en la science des nombres'. Which was begun continued and finished at Lyon on the Rhône in the year of grace 1484.
Explicit deo gracias.

The whole manuscript, which is on paper and is very easy to read, comprises 324 folios, i.e. 648 pages. In addition to the *Triparty*, and written by the same hand, in the same style, using the same notation and the same technique, we have a collection of 166 problems, followed by a section which applies algebraic methods to geometry. This begins on folio 211 and is entitled: 'How the study of numbers can be applied to the measures of geometry.'

Finally, there is a section on commercial arithmetic which ends on folio 324, with the words:

And thus ends our application of the study of numbers to affairs of commerce, and we come to the end of this book.
Explicit deo gracias.

Although the *Triparty* was not printed, its arithmetical passages were clearly the model for the 'Arithmetic newly composed by Maistre Estienne de la Roche, called Ville-franche, a citizen of Lyon on the Rhône, 1516'.

In the Sainte-Geneviève Library in Paris, there is a manuscript treatise on arithmetic by Jehan Adam, dated 1475, and, like the *Triparty*, written in French. Adam and

Chuquet both refer to a 'Maistre Berthelemy de Romans, doctor of theology, of the order of Predicant Friars, at Valence'. This French mathematician thus seems to have been famous and influential in his time but all other trace of him has now been lost.

In Venice, in 1494, Luca Pacioli published a bulky vernacular work entitled *Summa de Arithmetica, Geometria Proportioni et Proportionalita.* It is essentially an adaptation of the works of Leonardo of Pisa, rather like Chuquet's book.

Pacioli was born in about 1445. His family was poor and lived in the Umbrian town of Borgo San Sepolcro, the birthplace of the painter Piero della Francesca. Piero, himself a good mathematician, who took a serious interest in the theory of perspective, died on 13 October 1492. He left a manuscript of a Latin work on regular polyhedra, which without citing the original author, Pacioli incorporated, in Italian translation, into his own *De Divina Proportione*, published in Venice in 1509.*

In his youth, Pacioli (also known as Luca di Borgo) worked as a private tutor in Venice, and in 1470 he wrote a book on algebra which he dedicated to his employer. This work is now lost. It was while he was in Venice that Pacioli joined the Order of Franciscan Friars. In 1475 he was a public teacher of mathematics in Perugia. Later, he moved rapidly from city to city, visiting Florence, Pisa and Bologna. He died in Rome, some time in the year 1517. He was a friend of Leonardo da Vinci, who drew some magnificent illustrations for the *Divina Proportione*.

The only other fifteenth century scholar we shall consider in detail is Regiomontanus (Johann Müller), an astronomer whose work influenced the development of trigonometry.

---

* There was no 'copyright' at the time—and Piero's work is quite short (it forms less than one third of Pacioli's book). J.V.F.

In the fourteenth century the most important school of trigonometry had been at Oxford, where the outstanding scholars were John Mauduith and Richard Wallingford (the latter of whom died in 1335), but the fourteenth century mathematician who exerted most influence on Regiomontanus was the Provençal Rabbi Levi Ben Gerson (1288?–1344?) (also known as Leo de Balneolis or Leo Hebreus), who was attached to the Papal Court at Avignon.

At the beginning of the fifteenth century Johann von Gemunden (1380–1442) taught mathematics in Vienna, where his pupil, Georg Purbach (or Peuerbach) later took his master's place. Purbach was born near Linz, in a village from which he took his name, on 30 May 1423, and became astronomer to King Ladislas of Hungary before being appointed to the professorship in Vienna. Since the available Latin translations of the *Almagest* contained many errors, Purbach attempted to correct them, at least insofar as mathematics was concerned. 'So', Montucla writes [1], 'when the Great Cardinal Bessarion, who was interested in Astronomy, came to Vienna as Papal Legate he found it easy to persuade Purbach to learn Greek. But things were very different then from what they are now, when anyone can learn any language he pleases from a grammar book, without any contact with people who speak it. All the wisdom of Greece was to be found in Italy, where Greek scholars had taken refuge when they fled from the misfortunes of their native land. Bessarion persuaded Purbach to come back with him to Italy to learn the rudiments of Greek, and to bring with him his pupil Regiomontanus, who was no less eager than his master to learn the language. In 1461, when Purbach was about to set out, he unexpectedly fell ill and died, to the great regret of lovers of science.'

The German name of the astronomer whom we know by his adopted Latin name of Regiomontanus (he sometimes signs himself Joannes Germanus or Joannes de

Regiomonte) was in fact Johann Müller. He was born in 1436 in the small town of Königsberg (near Coburg) in Franconia.

Montucla writes: 'He was scarcely fourteen years old when, enchanted with mathematics, and particularly astronomy, he went to study under Purbach, who then enjoyed a high reputation. Regiomontanus was soon his favourite pupil, or, rather, his companion. During the ten years he spent with Purbach, that is until Purbach's death, he helped him in his various works ... Regiomontanus was to have accompanied Purbach to Italy, to learn Greek and thus be able to read the ancient sources in their original language. Death prevented Purbach from making the journey, and his pupil travelled alone to meet Cardinal Bessarion in Italy. There he learned Greek and made a translation of Ptolemy's *Almagest* from the original text. He also translated Theon's commentary on Ptolemy. It is hard to believe that one man could have made to many translations as he did, but these translations form only a small part of his writings. He made Latin translations of the *Sphaerica* of Menelaus, and the *Sphaerica* of Theodosius, as well as other treatises. Regiomontanus's interests went far beyond Astronomy: referring back to the original Greek he corrected the old translation of Archimedes made by James of Cremona [Montucla presumably means Gerard of Cremona]. He translated the *Conics* of Apollonius, the *On the Section of a Cylinder* of Serenus,* the *Pneumatica* of Heron, the *Music* and the *Optics* of Ptolemy, and his *Geography*, also the *Mechanics* of Aristotle, and many other works. His early death prevented him from publishing these various works, many of which are still to be found in manuscript form in the Library of Nuremberg, where all that could be found was painstakingly collected together.

---

* Probably fourth century AD, between Pappus and Theon (Heath).

'Regiomontanus did not confine himself to such work, which, though useful, cannot by its nature bring much honour. The number of his original works, several of them printed, is no less considerable. He continued the *Epitome*, or summary of the *Almagest*, which Purbach has left incomplete at his death and, on his deathbed, had asked his pupil to finish. Having fulfilled this duty, Regiomontanus then wrote a very clear and succinct commentary on Ptolemy, dealing with many of the astronomical problems connected with this theory. . . .

'He did not confine himself to astronomy: he was familiar with almost every other branch of mathematics as well, and there are few to which he did not contribute by his works: (1) He wrote a commentary on the books of Archimedes which had been neglected by Eutocius. (2) He defended Euclid's famous definition of proportional quantities against the imputations of Campanus and of the Arabs. (3) He disproved the Cardinal of Cusa's* claim to have squared the circle. (4) He wrote about weights, the building of aqueducts, burning glasses, and other subjects. . . . (5) He made considerable improvements to Trigonometry. This part of Regiomontanus's work is the one that does him greatest honour. . . .

'[His] work on trigonometry is contained in his treatise *De Triangulis*, in five books. It is a very thorough survey of both plane and spherical trigonometry. . . .

'Regiomontanus also excelled in engineering. Ramus attributes to him extraordinary devices even more wonderful than the best productions of modern engineers [1758]. For instance, he mentions a mechanical fly which set off from its master's hand, went round the table and returned to its starting point. He also speaks of an eagle which went to meet the Emperor and escorted him as far as the city gates.

---

* Nicolaus Cusanus.

But, as M. Weidler remarks, there is no confirming testimony from other authors, and it would be credulous to believe such tales. They must have arisen from Regiomontanus's great mechanical skill, and from the common people's love of anything marvellous. All that is actually known of Regiomontanus's mechanical inventions is that he and Walther made additions to the famous Nuremberg clock, one of the wonders of the time. Regiomontanus had also begun to make a machine he called an Astrarium, which was probably what we should now call a Planetarium. To judge from his description it must have been a very complicated machine.

'Regiomontanus suffered the same fate as his master: an early death put an end to all his projects by cutting him off in his prime. The visit to Italy that we mentioned lasted several years, after which he returned to Germany and in 1471 he settled in Nuremberg where he found a brilliant pupil in one of the citizens, Bernard Walther.... Regiomontanus stayed in Nuremberg, occupied with study and with observations, until 1475, when he returned to Rome at the invitation of Pope Sixtus IV, who asked him to help with a projected reform of the calendar. Regiomontanus was drawing up plans for a possible system of reform when he died. This great loss to mathematics took place in July 1476. All scholars mourned him; the Pope gave him a magnificent funeral and had him buried in the Pantheon. It is said that Regiomontanus's death was brought about by his criticism of George of Trebizond's translations of Ptolemy and Theon: the translator's sons could not abide this insult to their father's memory, and avenged it with poison. But although many authors have repeated this story I do not believe that it is founded on anything more than suspicion.'

7. Italian Triumphs

Quando che'l cubo con le cose appresso
Se agguaglia a qualche numero discreto.*

Tartaglia

The sixteenth century was the great age of elementary algebra. Italian mathematicians solved general cubic and quartic equations, while the mathematicians of Germany and France worked out a system of symbols which, after Vieta, was to become standard in all mathematical work.

We shall give brief descriptions of the lives of two great Italian mathematicians, and then go on to discuss the solution of cubic equations.

7.1. Cardan

Gerolamo Cardano, born in Pavia on 24 September 1501, was the illegitimate son of a lawyer, Facio Cardano.

> I learned from the scriveners that two hundred and sixty-nine years have passed since the birth of Aldo, my paternal grandfather, until this day, so it is believed that no family has been longer established in Italy than my father's. The same is true on my mother's side: for from Aluysius, my maternal grandfather, until this day one hundred and seventy years have passed.†

Here is a childhood memory of Cardan's which shows what was to be one of his particular qualities:

---

* 'When the cube and the things themselves add up to some discrete number.'

† *De Subtilitate*. 'This day' is the one on which Cardan is writing, not the date of his birth.

This happened from when I was four years old until I was seven: and every day from the second hour of the day until the fourth, or if I got up or woke up later, I thought I could see pictures by the end of the bed, shaped like little copper rings, showing trees, animals, men, cities, soldiers arrayed for battle, instruments of war, and of battle and other things, which went up and down one after the other. And of course I was much delighted with these visions, being a little child, and I looked at them attentively. Clare, my mother, and Margaret, my aunt, sometimes questioned me closely, asking whether I could see anything. As for me, although I was so young, I knew very well that it was a miraculous vision, so I assured them I could not see anything, fearing that if I told them of it the vision would leave me or some ill would befall me for having revealed such a secret.

Cardan studied first in his native Pavia and then went on to Padua, where he became Rector of the University. He received a degree in medicine in 1524 and established himself as a doctor in the little town of Sacco, got married, and then went to Milan, where he obtained a chair in mathematics at the Academia Palatina in 1534. However, he later lost this chair in a competition in which his adversary was another algebraist, Zuanne da Coi. In 1552 Cardan accepted the invitation to travel to Scotland as doctor to the Archbishop of Saint Andrews, John Hamilton, and spent about a year travelling through France, England and Scotland. The Archbishop, who suffered from asthma and dropsy, sent Cardan five hundred crowns to pay his travelling expenses. His salary was more than ten crowns per day and he retained the right to treat other patients. The Archbishop gave him two thousand three hundred crowns in money, and many presents. On his return journey, Cardan cast the horoscope of King Edward VI of England, then a young man of sixteen, very much weakened by an attack of scarlet fever that had been followed by an attack of smallpox. 'The king,' he said, 'was of rather less than average height. He had a pale face, with grey eyes and a solemn expression, decorous and handsome. He was unhealthy though not suffering from any particular disease, and one of his shoulder-blades stuck out a little'.

Cardan cast the young king's horoscope very carefully, read it attentively, and concluded, among other things, that he would certainly live beyond the average age, although after the age of 55 years 3 months and 17 days he was likely to suffer from various diseases. In the following July Edward VI died. Cardan explained that the length of the life of a weak person could not be predicted accurately by means of a single horoscope: to get an accurate result it would have been necessary to cast the horoscopes of all the people close to the king, and without these further details any horoscope that had been drawn up must be rather uncertain and predictions from it would therefore be correspondingly unreliable.

On his return to Italy, Cardan lived for a certain time in Pavia, where one of his sons was beheaded for the murder of his wife.

Cardan then obtained a chair at the University of Bologna, but on 14 October 1570 he was imprisoned on a charge of sorcery and was released only when he had promised that he would never teach again in the Papal States. In 1571 he went to live in Rome, where the Pope rewarded him for his medical skill by granting him a pension, which he received until his death in 1576.

The French historian de Thou writes the following in praise of Cardan:

'Jerome Cardan, of Milan, was a Mathematician and Doctor, who enjoyed a high reputation. His behaviour was strangely uneven and his life was marked by various adventures which he himself has described with a simplicity or a freedom not common among literary men. My readers will forgive me for not repeating the story here. A little before his death I saw him in Rome, dressed in a fashion quite different from the rest of the world. I often spoke with him, and I was exceedingly surprised to notice that, although his writings had made him famous, his person in no way corresponded with the reputation he had acquired in the world. I thus admired even more the excellent judgment of Julius Caesar Scaliger who, having made a study of Cardan's *De Subtilitate*, remarked that it contained many unevennesses, and that the writer, who in some passages seemed to rise above the level of human nature, in other passages was as irrational as a child. Cardan was very interested in Arithmetic and made many discoveries in this domain. He convinced many people of the correctness of judicial astrology, several times predicting things with more assurance and truth than one would expect from an adept at this art. But it was great folly and horrible impiety in him to decide to draw up the horoscope of Our Saviour, Jesus Christ, claiming that He who was in fact the ruler of the stars was subject to their laws. Cardan died in Rome on the 21st September, aged 75 years less 3 days, just as he had predicted, and it was said that he had starved himself so as to ensure that his prediction of his own death would be correct.'

Cardan says of himself: 'Plura scripsi quàm legi, plura docui quàm didici.'* His output was, in fact, considerable: the 1663 Lyon edition of his works takes up ten folio volumes. Cardan wrote about many different subjects. His most interesting mathematical works are the *Practica Arithmeticae generalis et mensurandi singularis* (1539) and the *Ars magna sive de regulis algebraicis* (1545). The latter describes Cardan's method of solving cubic equations, and it thus marks a very important date in the history of mathematics.

The following passage from *De Subtilitate* describes great men:

There are many who excel in the sciences and from them I have chosen twelve, leaving my readers to judge for themselves.

* 'I have written more than I have read, I have taught more to others than I have been taught myself.'

Let Archimedes be named as the first of the twelve, not only for the books he published, and for his mechanical inventions, by which he often overcame the might of Rome, as Plutarch says when he lists these marvellous inventions in the course of his account of the life of M. Marcellus: and we have listed others no less marvellous for which we have the testimony of Galen.* Archimedes was thus not only foremost but also inimitable in his inventions: and he did not disdain to praise those who followed the Graeculi and spoke a little Greek (which is why Cicero calls them 'Graeculi') and nor did he disdain to lie in a tomb placed among the ruins and the thorn bushes of the city of Syracuse.

My esteem for Ptolemy of Alexandria nearly equals that expressed for Archimedes, for Ptolemy's calculations on the stars are so clear and lucid that they will hold good for ever, and I have only ventured to mention the manner and subtlety of this divine work so that I can say that he worked it out. And as for other inventions, and many are known, there is nothing to match this one. I have good cause to hesitate in chosing the third man I shall praise, since there are many equally worthy: but let the third place be given to Aristotle of Stagira, the tutor of Alexander the Great when he was in Macedonia, for Aristotle gave admirable explanations of things natural and divine, and of dialectics, and also showed marvellous inventiveness in his descriptions of the lives and habits of animals. And although he wrote on all the sciences and his assertions have been questioned, nevertheless no error has been found in his works in the course of so long a time. By similar arguments Euclid, Scotus and John Suisset, who is commonly called the Calculator, all lay claim to the fourth place: but let Euclid come first, for his antiquity and for the sake of custom, which are two important reasons in the matter of praise. The firm constancy of his precepts, the books of the *Elements*, and their perfection, which is so absolute that we rightly refrain from comparing any other work with this one: for these two reasons the light of truth so shines from this work that only those seem capable of distinguishing truth from falsehood in difficult questions who have become familiar with the work of Euclid: he also wrote and composed many other works with similar subtlety and they too are used, though too little. He was born in Megara. John Scotus† comes next, from the very land which earned the name of subtle Doctor for its learning. John Suisset,‡ nicknamed the Calculator, came from close to Scotland, and all posterity has striven to explain one of his arguments which contradicts the experimental facts of mutual action: when he was very old he wept at not being able to understand his own arguments when he read them, or so it is said. So I do not suspect (and I have even written to this effect in my book on the immortality of the soul) that barbarians§ are inferior to us in mind or in

---

* Doctor (129–201 AD).

† John Duns Scotus, a philosopher of the thirteenth century.

‡ Richard Swineshead, professor at Oxford in the fourteenth century.

§ Probably used in the Greek sense of 'foreigners'. J.V.F.

understanding, since England, in the Far North, separated from the rest of the world, has produced two such brilliant men.

The seventh place in my praise must be given to Apollonius of Perga, who was almost the same in age as Archimedes. He composed eight excellent books on the Elements of the pyramid, though the first four have so far been treated so ill by the translator that they can hardly be said to have been published.

Archytas of Tarentum follows closely: but he lived too long ago for an Italian to be given his due of praise among such noble men. It is said he made a wooden dove which could fly, and he gave the first correct method of constructing two lines in continuous proportion between two given lines, a method Eutocius published among ten other incorrect ones.

Mahomet, son of Moses the Arab,* follows Archytas, for he invented the art of Algebra, and because of his invention he was given the name Algebras, from the name of the art.

Alchindus† is the tenth in my praise. He was an Arab who published many books, which are mentioned by Averroes, and there is one little book on the ratio of six quantities, which I shall try to have printed, since nothing can be more ingenious than its arguments.

Then comes Heber the Spaniard,‡ excellent in his invention: and just as Ptolemy seeks hard to find the sixth to follow five quantities, so Heber with these same quantities seeks the fourth to follow three others. He gave many better descriptions of the sky or of the air: so you may now easily understand that great heat causes less harm to the mind than great cold does.

Galen is the last in subtlety; but the most excellent in art,§ method, taking a pulse and dissections: nevertheless he is so wordy, and so tedious in his love of contradicting others, that if he were someone else you would find him intolerable: for the rest, his writings are the great boast and treasured possession of the arts, which those of our own time have tried to restore.

Among these excellent and noble men we must also number Vitruvius,‖ although he was a practical engineer, who, if he had written of his own inventions instead of those of others, could have been numbered among the first. Some achievements of each author are outstanding: in Archimedes the equality of the sphere with the cylinder, inscribed on his tomb: in

---

* Al-Khovarizmi.

† Al-Kindi (Abû Yûsuf Ya qûb ben Ishâq ibn al-Sabbâh) who died about 873. He was born at Bassorah and was a polymath. A large part of his writings were translated by Gerard of Cremona.

‡ Abu Muhammad Gabir Ben Aflah, called Geber in Latin, a twelfth century astronomer who lived in Seville.

§ Medicine was taught in the Faculty of Arts. J.V.F.

‖ Roman architect, first century BC.

Euclid the order: in Aristotle the context; in Galen the style of discourse is excellent. Some few other authors also have excellent qualities and merits, though they are not superior in subtlety. Who is not amazed by the emotions aroused by Homer, the dignity of Virgil, the sympathy and richness of Cicero, the quality of the rhetoric: and the appropriate use of imagery in Quintilian? There is thus not one source of subtlety, but many, for which authors are famous and praised. Aristotle's subtlety comes from his understanding, which was imitated by Theophrastus* and John Scotus: the subtlety of Archimedes is that of understanding and imagination: there is subtlety through imagination, like that of John the Calculator: through the senses and imagination like that of Euclid: the subtlety which comes from reason, like that of Ptolemy: which comes from judgement, like that of Algebras: the subtlety which is of the senses and of experience, like that of Vitruvius, whom Heron emulated with many excellent though not useful inventions. Vitruvius is excellent on the subject of clocks, either water clocks, or ones where the dial turns according to the zodiac or in elevation, and also in discussing the common properties of hollow or concave objects and of flat and round ones.

7.2. Tartaglia

Niccolo Tartaglia was born around 1500, into a very humble family in Brescia. He became an orphan at six years old, and was seriously wounded during the sack of Brescia on 19 February 1512. For a long time he spoke with a stutter, and this gave him the nickname Tartaglia (stutterer).

He had no teachers but, despite heavy family responsibilities, succeeded in gaining a wide scientific knowledge. He taught mathematics in Verona, Plaisance, Venice, Brescia, and then again in Venice, where he died on the night of the 13/14 December 1557.

He himself tells us of the difficulties he had with the authorities of Brescia in about 1548, when he was a public teacher in his native city for eighteen months. He had come from Venice with all his family, having been promised an honourable salary, but, despite his pleas, he was not paid, and had to leave again for Venice without receiving one penny of what was owing to him.

---

* Naturalist, pupil of Aristotle (372–288 BC).

Tartaglia's most important work is his *General trattato di numeri et misure*, about one thousand five hundred pages long, which appeared between 1556 and 1560. It contains treatises on arithmetic, practical geometry and algebra, and a good Italian translation of Archimedes's *On the sphere and cylinder*, as well as a treatise on the division of figures which is in the tradition of Heron and Leonardo of Pisa. The section on algebra does not contain a description of the method of solving cubic equations although it was this work that had made Tartaglia famous.

Tartaglia was more than willing to lay claim to other people's scientific work, but his own works, mainly written in the Italian dialect of the Po valley, bear witness to great knowledge and to a real genius for mathematics. Moreover, either by simply appropriating them, or sometimes by citing their real authors, Tartaglia helped to put back into circulation various thirteenth-century ideas on mechanics and also some works of Archimedes, thus rendering a considerable service to future scholars.

Tartaglia tells us about his early misfortunes in a passage of his work *Quesiti et invenzione diverse* (1546):

P.—Tell me, do you remember having known me when I lived in Brescia?

N.—Yes, I remember you, though I was very young at the time. And, to show I really do remember you, your Honour lived in the part of the city between the Carmini and Santo Christofolo, or new Santa Chiara.

P.—You're right. But, tell me, what was your father's name?

N.—My father was called Michael, and, since nature had been as mean in giving him height as fortune was in letting him participate in her benefits, he was called Micheletto.

P.—Certainly, if nature was mean about giving your father a reasonable height, she was not particularly liberal with you.

N.—I'm glad of that, because being so short proves that I'm really his son, and although he left us (myself, my brother and my two sisters) almost nothing except happy memories, it's enough for me to have heard from many people who knew him, and were friendly with him, that he was a good man. That pleases me a lot more than if he had left me many goods and a bad name.

P.—What was your father's trade?

N.—My father owned a horse, on which he went messages for the citizens of Brescia, I mean that he took letters from the noblemen of Brescia to Bergamo, to Cremona, to Verona and to other places like that.

P.—What was his family name?

N.—Indeed, I don't know. I remember nothing about his family, not even the name, except that, since he was small, I have heard him called Micheletto the rider. Perhaps he had another name, but not one I know. You see, my father died when I was only about six years old, so I was left with a brother a little older than myself and a sister a little younger, together with our widowed mother, quite without any kind of fortune. We were soon completely bankrupt—which would be a very long story to tell, but it gave me plenty of other things to think about rather than my father's family name.

P.—Since you don't know your father's family name, how comes it that you are called Niccolo Tartaglia?

N.—I can tell you how that happened. The French sacked the city of Brescia and in the sack Signor Andrea Gritti was captured (Gritti of noble memory who was Provisore of the city at the time) and he was carried off to France and his house was completely plundered (although there wasn't much in it). I had taken refuge in the Cathedral with my mother, my sister and several other men and women from our part of the city, hoping that in such a place we should at least be safe from personal injury. But the hope was vain, and there, in the church, in front of my mother, I was given five mortal wounds, three on my head (each of them exposing the brain) and two on my face. If my beard did not hide the scars, I should look like a monster. One wound went right across my mouth and my teeth, breaking my jaw and palate in two. This wound not only stopped me from talking, except with my throat the way that magpies do, but I could not even eat because I could not move my lips nor my jaws, since they were broken along with my teeth as I've said, and things were so bad that I had to be fed with liquid, and even that was very difficult. Worst of all, my mother, who could not afford to buy ointment nor to call the doctor, had to look after me herself, and she could not use any ointments but could only frequently wash my wounds, following the example of dogs who, when they are wounded, just lick their wounds with their tongues. She nursed me back to health after several months, and, to return to our story, once I was cured of these wounds there was a time when I could still hardly talk and always stuttered, because of the wound across my mouth and my teeth (which are still not quite firm) and the children my own age that I talked to gave me the nickname Tartaglia (the stutterer). Since this surname lasted a long time, as a memento of my defect, I decided to call myself Niccolo Tartaglia.

P.—How old were you at the time?

N.—About twelve.

P.—It was certainly very cruel to wound a child of that age. Your strange name surprised me, I admit, because I couldn't remember having heard such a family name in Brescia.

N.—I have told you exactly what happened, your Excellency.

P.—Who was your tutor?

N.—Before my father's death, I was sent to school for a few months to learn to read, but I was very young at the time, between five and six years old, and I don't remember the name of my teacher. When I was about fourteen I chose to go to the writing school of a certain Master Francesco, and I began to write the alphabet just as far as the letter k, using commercial characters.

P.—Why only as far as k and no further?

N.—Because the terms of the contract with that particular teacher were to pay a third in advance, and another third when one knew how to write the alphabet as far as the letter k, and the rest when one knew how to write the whole alphabet. Since at the time, I did not have the money to pay him, and since I wanted to learn, I was trying to get hold of one of his complete alphabets and the model letters he'd written out, and after that I didn't go back any more. Since then I've learned on my own. I've had no other tutor, but only the constant company of a daughter of poverty called industry. I've always worked from the books of men who are dead. Since I was twenty I've always been burdened with numerous family responsibilities, and now cruel death has left me almost completely alone.

In Chapter 6, Vol. 2, we shall outline the method of solving cubic equations described by Hudde in the seventeenth century. It is, in principle, identical with that discovered by the Italians in the sixteenth century.

It seems to have been invented by Scipione dal Ferro, who was born in Bologna on 6 February, 1465 and died in the same city at the end of October or the beginning of November in the year 1526. He taught arithmetic in his native city from December 1496 onwards and lived in Venice for several months during the last year of his life. His son-in-law, Annibale della Nave, succeeded him in his chair, and inherited his papers (now lost) which contained a description of his method of solving cubic equations. These papers were never published but several mathematicians had access to them.

In 1530, Zuanne da Coi set Tartaglia several third-degree problems. We do not know whether Tartaglia had had wind of Scipione dal Ferro's discovery or whether he had worked out a method for himself, but a few years later Tartaglia certainly knew how to solve equations of the form $x^3 + px + q = 0$ which have one real root.

When Cardan heard about this he at once asked Tartaglia to explain the method to him. In March 1539, when they met in Milan, Tartaglia finally consented to tell his secret, which he expressed in the following lines of verse:

Quando che'l cubo con le cose appresso
Se agguaglia a qualche numero discreto:
Trovati dui altre differenti in esso.
Dapoi terrai, questo per consueto,
Che'l loro produtto, sempre sia eguale
Al terzo cubo delle cose netto;
El residuo poi suo generale,
Delle lor latti cubi, ben sottratti
Varrà la tua cosa principale.
In el secondo, de cotesti atti;
Quando che'l cubo restasse lui solo,
Tu osserverai quest' altri contratti,
Del numer farai due, tal part'a volo,
Che l'una, in l'altra, si produca schietto,
El terzo cubo delle cose in stolo;
Delle qual poi, per commun precetto,
Torrai li lati cubi, insieme gionti,
Et cotal somma, sarà il tuo concetto;
El terzo, poi di questi nostri conti,
Se solve nol secondo, se ben guardi
Che per natura son quasi congionti.
Questi trovai, et non con passi tardi
Nel mille cinquecent'e quattro e trenta;
Con fondamenti ben saldi e gagliardi
Nella Città del mar intorno centa.

The first nine lines can be roughly translated:

When the cube and the things themselves add up to some discrete number, take two others different from the first, choosing them so that their product is always equal to the cube of one third of the third thing, and the difference between their cube roots will give you the main thing.

We are required to solve $x^3 + ax = b$, where a and b are positive. We find numbers u and v such that $u - v = b$ and

$uv = (a/3)^3$. Then the root of the equation is

$$= \sqrt[3]{u} - \sqrt[3]{v}.$$

The following nine lines concern the equation

$$x^3 = ax + b,$$

and the next three leave it to the reader to generalize the argument. The last four lines date the method to 1534, and say it was invented in the City surrounded by the sea, that is, in Venice.

Cardan overcame his initial difficulties in applying this method and went on to do extensive work on cubic equations, even coming close to recognizing the existence of what we should, in modern terminology, call complex numbers.

We cannot follow him over this new ground here. His pupil, Ludovico Ferrari (who was born in Bologna on 2 February 1522, and died in October 1565), completed Cardan's work by solving quartic equations.

The crowning achievement of the Italian school was the *Algebra* of Rafael Bombelli. Nothing very much is known about Bombelli's life except that he was an engineer. His treatise, which appeared in 1572, contained the first serious discussion of the properties of complex numbers. These numbers were later to play a very important part in mathematics.

8. The Father of Modern Mathematics: François Viète

Franciscus ille Vieta Fontenaensis, cujus
mirae in Mathematicis lucubrationes Veteri
Geometriae felices praestitere suppetias.*
Fermat

From the sixteenth century onwards, a great deal of sophisticated mathematical work was done in many different schools and in many different countries, so we shall be obliged to be selective in the historical account we give in the following chapters.

Our last chapter was devoted to Italian algebraists. The present one concentrates entirely on French mathematicians, and ends with a description of one of the greatest geometers of all time, Franciscus Vieta (François Viète).

At the beginning of the sixteenth century the study of mathematics at the University of Paris owed much to the enthusiastic and diligent work of Jacques Lefèvre who was born in Étaples about 1455 and died at Nérac in 1536, 'a little man of very low birth' according to Bayle. He is important chiefly for his editions of the works of Euclid, Jordanus Nemorarius and John Sacrobosco, and for the commentaries he wrote on these works.

Oronce Finé, who was born at Briançon in 1494 and died in Paris on the 6 October 1555, taught mathematics at the Collège Royal from 1530 until 1555. One of his first

---

* François Viète of Fontenay whose marvellous work in mathematics so happily extended ancient geometry.

works, written in 1519, was a new edition of the *Arithmetica Theoretica et Practica* of Juan Martinez Siliceus.*

Finé was a popularizer rather than an original mathematician, and was also excellent cartographer. The most notable of his works were the *Arithmetica Pratica*, printed in Paris in 1530 and reprinted in 1544, and the *Protomathesis* (in fifteen books), published in 1532. He is famous as one of those who tried to square the circle, and as one of the first mathematicians to use decimal numbers.

Before turning to the true algebraists, we shall refer briefly to Ramus (Pierre de la Ramée) who was one of the most brilliant figures of the University of Paris.

He was born in 1515, and since his family was poor he had to take a job as a servant in the Collège de Navarre in order to be able to study there.

'He had a good figure and a handsome face. He was of a vigorous disposition and was a tireless worker. He never slept on anything except straw, from childhood until old age. He usually rose at cock-crow. Since he spent all day reading, writing and thinking, he only took a light meal in the morning, so that his spirit should be more free. In the evening he ate rather more, and after supper he went for a walk for two or three hours, or talked with his friends. His usual food was boiled meat, and he never drank wine until quite late in his life.

'His life was that of a continent celibate. His health was good and he recovered from all his illnesses not by using remedies but by sobriety, abstinence, exercise and, above all, by playing tennis, which was his usual form of amusement.

'Ramus was one of those who maintained that French spelling ought to agree with pronunciation.' [1].

---

* Juan Martinez Siliceus (the pebble, perhaps because of his rough countenance), 1477?–1557, taught at the University of Paris for some time. His work, published in Paris in 1514, is not without merit. He was tutor to Philip II and when he died he had become Archbishop of Toledo and a Cardinal.

Pierre de la Ramée made vigorous attacks on Aristotle's philosophy and on Euclid's geometry, and while it is not our concern here to criticize him as a philosopher, we must admit that his diatribes against Euclid are sometimes so weak and so naive as to be ridiculous. He was, however, the first French historian of mathematics, and can also be considered as the first proponent of the French system of elementary mathematical teaching, a system that was to be adopted for the following three centuries. Moreover, he founded a chair of Mathematics at the Collège Royal, a chair which bore his name and was filled by a competition held every three years. Roberval was to be its most illustrious occupant.

Ramus, like Jacques Lefèvre before him, was a Protestant. He was killed by a student mob on the third night of the massacre of St. Bartholomew's Eve in 1572.

Jean Borrel, or Buteo, or Buteon, was born at Charpey in the Drôme in 1492. He joined the order of St. Anthony of Vienne and when he was thirty years old his superiors allowed him to go to Paris to complete his education. The licence given him for printing his works, dated 22 January 1553, calls him 'Maitre Jehan Buteo, Commander of the Order of Saint Anthony'. He administered the estate and the château of Balan, which were a few miles from the Abbey, and he six times saw the Abbey taken and sacked by the Calvinists. He is said to have died of a broken heart, far from his books, in 1572, in a retreat at Canar near Saint-Nazaire-en-Royans (Drôme) or in a hamlet of the commune of Miribel-les-Echelles (Isère).

His main work was an original treatise on algebra called *Logistica, quae et Arithmetica vulgo dicitur, in libros quinque digesta*, which was published in Lyon in 1559. In it he is particularly concerned with the elimination of unknowns between several linear equations. Guillaume Gosselin was later to improve upon his methods.

According to de Thou, 'Buteo also made several inventions and was particularly ingenious at devising musical instruments and new machines, in which he gave great evidence of his industry.'

He was the first critical historian to notice that the proofs Euclid gives in his *Elements* must be attributed to Euclid himself and not to his much later editor, Theon of Alexandria.

We have mentioned Guillaume Gosselin, of Caen, but little is known about him: it is not even known when he was born or when he died. His French translation of the first two parts of Tartaglia's *Arithmetica*, published in Paris in 1578, is dedicated 'To the very illustrious and virtuous Princess Marguerite of France, Queen of Navarre', and the letter of dedication is dated 'From Paris at the Collège de Cambray, this second day of November 1577.' Gosselin undertook to write a commentary on the work of Diophantus and Cardinal Duperron (Jacques Davy, 1556–1618) lent him a manuscript of the thirteen books of the *Arithmetica*. But Gosselin soon afterwards died of the plague* and both the commentary and the manuscript were lost. Gosselin's main work is a treatise on algebra:

> *Gulielmi Gosselini Cadomensis Bellocassii de Arte magna, seu de occulta parte numerorum, quae et Algebra et Amulcabala vulgo dicitur, Libri Quatuor. In quibus explicantur aequationes Diophanti, Regulae quantitatis simplicis et Quantitatis surdae. Ad Reverendissimum in Christo Patrem Reginaldum Bealneum,† Mandensem Episcopum, Illustrissimi Ducis Alenconii Cancellarium, Comitem Gevodanum, atque in sanctiori et interiori consilio Consiliarium, Parisiis, apud Aegidium Beys, via Jacobaea, ad Insigne Lilii albi, MDLXXVII (in-8°).*

Jacques Peletier who was born at Le Mans on 25 July 1517, came from a good middle-class family: of his two

---

* Perhaps during the epidemic which broke out in Paris in June 1580, and killed 40 000 people in five months.

† Renaud de Beaune, born in Tours in 1527, Bishop of Mende from 1568 to 1581, appointed Archbishop of Bourges in 1581, and Archbishop of Sens in 1596. He died in 1606.

elder brothers one, Julien, was an Advocate at the Parliament, and the other Jean, taught at the Collège de Navarre where he eventually became a Grand Master. Jacques entered this college at the age of five and was a brilliant pupil. He studied Law without much enthusiasm, and in about 1537 joined his brother, then a teacher. He was active both in literature and in science. He was presented to Marguerite de Valois, the sister of François Ier, and was friendly with the poet Clément Marot (1496–1544). In 1539 he met the young Théodore de Bèze*, to whom he was later to dedicate his *Arithmetic*.

In 1540, when Marguerite was compelled to disperse her circle of intellectuals and courtiers, Peletier returned to Maine to become secretary to René du Bellay, the Bishop of Le Mans. In 1543, at the funeral of Guillaume du Bellay,† he met Ronsard,‡ on whom he was to exert considerable influence. Peletier returned to Paris in 1544, but left again in 1548, and went to live in Bordeaux. He did not like Bordeaux, so he moved to Poitiers, where he published some short works on orthography and the four books of his *Arithmetic* (published in 1549).

Peletier left Poitiers in 1552 to live in Bordeaux, Béziers and Lyon, occupying his time with Medicine and Mathematics.

It was in Lyon that he published several of his lyrical works, his *Algebra* (1554), which is in French and uses curious spelling, as well as his Latin edition of six books of Euclid (1557).

After a short stay in Italy, Peletier lived in Paris from 1557 to 1569, and took a degree in medicine at the University in 1560. He left Paris for Savoy, but returned in

---

* Protestant writer and theologian (1519–1605): a pupil of Calvin. J.V.F.

† A general of François 1er, cousin to Joachim du Bellay (1522–60) the poet, one of the Pléïade. J.V.F.

‡ Poet (1524–85), also one of the Pléïade. J.V.F.

1572. He later went to Bordeaux to escape the wars, and spent seven years of study there, during which time he published: *De l'usage de Gèométrie, par Jacques Peletier, Médecin et Mathématicien. A très illustre Seigneur Messire Albert Degondy Comte de Rets. A Paris chez Gilles Gourbin, à l'enseigne de Espérance, devant le Collège de Cambray MDLXXIII.**

In 1579 Peletier returned from Poitiers to Paris, perhaps to take up an appointment as Head of the Collège du Mans in the rue de Reims on the Montagne Sainte-Geneviève. He died in July 1582.

As an indication of his quality as a poet, here is a short extract from his Oeuvres Poétiques of 1547:

Au reverendissime cardinal Dubellay

> Le Clair Soleil aux estoilles depart
> De sa splendeur, sans qu'ell'en diminue:
> Maint beau ruisseau d'une fontaine part,
> Sans que la source en rien discontinue:
> Sus cest égard ma voye j'ay tenue
> Vers vous, auquel les lettrez ont recours,
> Pour impétrer faveur, grace et secours:
> Afin qu'un jour je vous nomme à voix claire
> La source vive où commence mon cours,
> Et le Soleil qui à ma nuit esclaire.†

We cannot form an opinion of his qualities as a medical man, but as for the mathematician, Montaigne‡ says:

---

* *The use of Geometry by Jacques Peletier, Doctor and Mathematician. To the very illustrious Lord Messire Albert Degondy Comte de Rets. In Paris at Gilles Gourbin, at the Sign of Hope, in front of the Collège de Cambray MDLXXIII.*

† To the Most Reverend Cardinal Dubellay
The bright sun shares its light among the stars without diminishing its own glory. Many a fair stream flows from a fountain without drying up the spring. In the same way I have looked to you, to whom scholars have recourse for favours, grace and help, so that one day I can clearly call you the living spring from which my being flows, and the sun that shines through my darkness.

‡ 1533–92. Chiefly famous for his essays. J.V.F.

'These however [facts and reason] often conflict with one another, and I have been told that in Geometry (which seems to be the most certain among the sciences) there are irrefutable proofs which run contrary to what we in fact experience: for instance, Jacques Peletier was telling me one day when he visited me that he had found two lines which were inclined towards each other as if they would meet, but he had shown that it was absolutely impossible for them to do so even if they were extended to infinity. . . .'

Montaigne's story shows us Peletier at his most characteristic: ahead of his time in his constant preoccupation with the infinitely great and the infinitely small. This same preoccupation led him to quarrel with Clavius about the angle of contingency, a quarrel which was important for its effect upon Vieta.

François Viète was born in 1540 at Fontenay-le-Comte. His family surname was Viète (or Viette, as it is still spelled by the Western branch of his family and as it was spelled by many of his contempories) but in English it is usual to use the Latin form 'Vieta'.

Vieta was the son of an Attorney, and was educated in his home town at the college run by the Grey Friars.

He studied Law at Poitiers, and was called to the Bar at Fontenay in 1560, but in 1564 he abandoned the legal profession to enter the service of the house of Soubise. The head of the family, Jean Larchevêque de Parthenay, Seigneur of Soubise, was one of the military leaders of the Huguenots of Poitou. He had taken over the command of Lyon from the Baron des Adrets, and in 1562–3 had saved the city when it was besieged by the Duke of Nemours, despite the fact that in an attempt to persuade him to yield the Catholics had threatened to kill his wife and daughter before his eyes. One of Vieta's first tasks in his new employment was to draw up a 'Description of what happened at Lyon while M. de Soubise was in command there'. Théodore de Bèze included this in his *Ecclesiastical History of the Reformed Churches* (*Histoire ecclésiastique des Eglises reformées*). Vieta also wrote some 'Memoirs of the life of Jean de Parthenay' and a 'Genealogy of the House of Parthenay Lusignan,' both of which remained in manuscript.

Jean's daughter, Catherine de Parthenay, born in 1552, was twelve years old when Vieta entered her father's service to become her tutor. One of the textbooks he wrote for her, survives. It is called *Principles of Cosmography* (*Principes de Cosmographie*) and was published in 1637, being reprinted in 1643, 1647 and 1661. In 1591 Vieta dedicated his *Ars Analytica* to his pupil. In 1568 Catherine married the Baron de Quellenec, who was killed in the massacre of St. Bartholomew's Eve (1572). She later married René de Rohan, who died in 1585. Catherine played an important part in the defence of La Rochelle against Richelieu, and then withdrew to live with her family on the Soubise family estate in the wooded district of the Vendée, where she died in 1631.

After Catherine's first marriage and the death of Jean de Parthenay, which took place in 1567, Vieta went to La Rochelle as secretary to Jean's widow, Antoinette d'Aubeterre, Dame of Soubise.

In 1571 Vieta was an Advocate at the Parliament of Paris and in 1573 he was a Councillor at the Parliament of Rennes under the protection of the Rohan family. In 1576 King Henri III entrusted him with carrying out special missions and services. In 1580 he became Referendary to the King's Household. He was also a member of the Privy Council, but the influence of the Sainte-Ligue led to his being relieved of his duties from the end of 1584 until April 1589, when Henri III broke with the Guise faction.

Vieta married Juliette Leclère, and she bore him one child, a daughter who died without issue. He owned the estate of

Bigotière at Mervent near Fontenay-le-Comte. He died on 23 February 1603.

Vieta's official duties occupied most of his time and there were only two periods of relative leisure in his life: from 1564 to 1568, and then from 1584 to 1589. It was during these two periods that he concerned himself with mathematics.

He turned first to astronomy and trigonometry: his *Harmonicon coeleste* dates from the first period of leisure (1564–8). This work was never printed. On the 28 December 1627, Jacques Dupuy wrote to Peiresc about Alleaume's papers, which included several manuscripts by Vieta:

> My brother having heard of a book called *Harmonicon coeleste* sent M. de Lomenie to ask for it and he was given M. Vieta's own sketch for it. The sketch is quite clear in many places but equally confused in others. He has passed it on to us and we shall keep it in case we need it. There was a fair copy made of the complete work in its final form, but this has no doubt been taken. However, if any plagiarist were to attempt to lay claim to the work, the sketch that we have would be sufficient to establish the true authorship.

This autographed manuscript was lent to Mersenne by Pierre Dupuy. It was stolen from him, and the Dupuy brothers only got it back in 1647. Copies of it are now preserved in Paris and Rome.

It was also in the period 1564–8 that Vieta began to write his *Canon mathematicus*, which was printed over the years 1571 to 1579. This work, in which Vieta calculates numerical values of the circular functions (sine, cosine etc.), is the first to make systematic use of decimal numbers, which Stevin's work was soon to make popular.

In his second period of enforced leisure Vieta drew up the broad outlines of his *Ars analytica*.

Thanks to Père Mersenne and to Schooten, the Elsevier Press was able to publish some of Vieta's works in 1646, but the edition was by no means complete. On the other hand, manuscripts of his work were in circulation among scholars: Vieta himself gave some to Marino Ghetaldi when he visited Paris, and when Vieta died Alleaume inherited several of them, some of which he had published. Others were published shortly after his death. The following works appeared in this way:

In artem analyticam isagoge. Published in Tours (1591) and then in Paris (1624).

Supplementum geometriae. Published in Tours, 1593.

Zeteticorum libri quinque. Published in Tours, 1593.

Variorum de rebus mathematicis responsorum libri VIII. Published in Tours, 1593.

Ad problema, quod omnibus mathematicis totius orbis construendum proposuit Adrianus Romanus, responsum. Published in Paris, 1595.

Apollonius Gallus seu exsuscitata Apollonii Pergaei περι επαφων geometria. Published in Paris, 1600.

De numerosa potestatum ad exegesin resolutione. Published by Marino Ghetaldi, Paris, 1600.

De aequationum recognitione et emendatione et Analytica angularium sectionium. Published by A. Anderson, Paris, 1615.

De Thou says of Vieta:

'François Viète, born at Fontenay in Poitou, was a profound and original thinker who understood the most secret mysteries of the most abstruse sciences and easily succeeded in all the projects that an intelligent man could conceive and carry out. But among his various occupations, and the wealth of affairs with which his vast and tireless spirit was always occupied, he more particularly turned his attention to mathematics, and his excellence was such that all the results obtained by the Ancients, of which time has deprived us by destroying their writings, all these he rediscovered for himself and recalled them to men's minds, sometimes even adding further results of his own.

'His powers of concentration were such that he often remained three days together in his study without eating and even without sleeping, except by occasionally resting his head on his hand to refresh himself with a few moments of slumber.

'He wrote several books but copies of them are exceedingly rare because since they were printed at his own expense he received all the copies and, being very generous, he gave copies to all who were interested. Apart from his original works, he left many others which shed light on the arts and served to revive the memory of Ancient authors and since Pierre Alleaume of Orleans* had helped him in his work, Vieta's heirs gave him the manuscripts. It is from this rich collection that Alleaume, the Scot Alexander Anderson† and several others have taken the material for many of the treatises that they have published (to the great admiration of all lovers of mathematics). These works are a living memory to the glory of this great man.'

'Adrian Romanus‡ proposed a problem to all the mathematicians of Europe and Vieta, who was the first to solve it, sent his solution to Romanus with corrections and a

---

* Pierre Alleaume, Advocate at the Parliament of Paris, a friend of Vieta, whose heirs passed many of Vieta's manuscripts on to him. He himself left them to his son Jacques, Vieta's pupil. It seems that De Thou has confused the father with the son. Jacques Alleaume was an engineer to Sully in 1613 (when he was Grand Master of artillery and fortification) and was later a friend of Snellius and of Beeckman (1588–1637), Descartes's friend, when he (Alleaume) served at Breda under Prince Maurice of Orange, who died in 1627.

† Vieta's pupil, Alexander Anderson, probably a Protestant, was born in 1582 and died some time after 1621. Vieta had also given several of his manuscripts to Marino Ghetaldi of Ragusa (1566–1627).

‡ Adriaan van Roomen or Adrianus Romanus (1561–1615) was born at Louvain. He taught at Louvain and then at Würzburg. He was a talented mathematician.

proof, together with *Apollonius Gallus*. Romanus was so impressed by Vieta's knowledge that he set out from Würtzburg in Franconia, where he had been living since he had left Louvain, and travelled to France to make Vieta's acquaintance. When he arrived in Paris he found that Vieta had gone to Poitou for the sake of his health. Romanus followed him there, although it meant a journey of about a hundred leagues, and when he finally had the pleasure of meeting Vieta he consulted him at length about all the difficulties he had encountered. Such was his admiration for this extraordinary man that he admitted that what he had found exceeded even what he had imagined. Romanus stayed with Vieta for a month and left him only with great regret. Vieta, wishing to recognize the honour Romanus had done him in undertaking so long a journey to visit him, arranged to pay for his guest's return journey through France.

'Moreover, Vieta's essay on Apollonius was held in such high esteem that it inspired Marino Getaldi of Ragusa, a very excellent mathematician, to write a book called *Apollonius Redivivus*. This was published seven years later, with the *Apollonius Gallus* included as an appendix.

'It displeased me much that Scaliger* attacked Vieta with so much bitterness on the subject of Cyclometres. But at the time this generous man did not recognize the full merit of his adversary, and thus could not forbear to show resentment when he was corrected by him although he had not in fact fully examined the logic of his own argument. Later he corrected his mistake and with a praiseworthy frankness withdrew his attack. From then on he always had a secret admiration for Vieta.

---

*Joseph Scaliger (Agen 1540, Leyden 1609), a scholar. Known in mathematics as one of those who tried to square the circle. A friend of De Thou.

'A little before his death Vieta, realizing that the Lilian calendar* had several deficiencies which had already been pointed out by other people, succeeded in designing a reformed version which was suitable for use by the Catholic church and adapted to its festivals and rituals. This was printed in 1600, and he gave it to Cardinal Aldobrandini at Lyon when he arrived there as the Papal envoy sent to settle the dispute between the King and the Duke of Savoy. But Vieta's enterprise met with ill success, as I had warned him it would when he told me about it before he set out. Many people had striven to introduce a reformed calendar in the various states of Christian princes, where such a reform was, finally, accepted after a great deal of effort, but such people make it a rule of statecraft not to admit to having made errors, or even to admit it is possible that they should ever have been in error, and they were therefore not willing to accept a change which would make it clear that they had in fact made mistakes in the past.

'When peace had been made, Cardinal Aldobrandini returned to Rome, and Christopher Clavius, who had already been occupied with the work of Lilius, rejected the correction Vieta had proposed to the Cardinal. Vieta then wrote to this famous mathematician, complaining bitterly at his behaviour, and it seems that if he had not died soon afterwards the dispute would not have ended where it did; for those who were not afraid to attack so redoubtable opponent after his death would not have attacked him with impunity during his lifetime.

---

* The Gregorian calendar now in use. Pope Gregory XIII decided to reform the Julian calendar. The reform was carried out in Rome in 1582 and successively adopted by various States. It had been worked out by a committee of several priests and several scholars, among them the Jesuit Father Clavius, professor of mathematics at Rome (1537–1612). The committee was working on a project for reform devised by Aloisius Lilius, a Veronese doctor and astronomer of Verona who died in 1576.

'Before this dispute stirred up bitterness between them, Vieta had made it clear that he considered Clavius an excellent writer on the elements of mathematics and one who gave lucid and convincing explanations of many obscure passages in original works, although he tended to write as if he had only just beome acquainted with his subject matter and made no original contributions to it. He merely copied from the works of others, without acknowledging his debt to the original authors, despite the fact that his own contribution had merely been to collect, arrange and explain what he had found scattered in various places in the books he used, which were, indeed, not always models of order and clarity.*

'What I am about to add is not important, according to Vieta's own view, though anybody else would consider it so. The State of Spain consists of several geographically separate parts, and the various communications that have to pass between them and, are, for the sake of secrecy, written in various ciphers. When the Spaniards want to change these ciphers they can do so only some considerable time after taking the original decision, because they have to give advance warning of the change to the Viceroys of the Indies.

'At the time of the disturbances of the Sainte-Ligue the cipher the Spaniards used consisted of more than 500 different characters, and although many exceedingly long letters containing explanations of Spanish plans had been intercepted, all attempts to decipher them had failed, because so many signs were involved. The King had these letters be sent to Vieta, who had no difficulty in understanding them, nor in understanding all the later ones. This so amazed the Spaniards, and so upset their plans for

---

* Vieta's judgment on Clavius, as reported by de Thou, is on the whole fairly just, but nevertheless rather too severe. Clavius's textbooks contributed much to the progress of mathematics.

two whole years, that they proclaimed in Rome, and everywhere else, that the King must have used magic to learn their cipher.'

As we shall see in Vol. 2, Chapter 2, Vieta invented modern mathematical symbolism. Mathematics was, for him, essentially synonomous with Analysis, which could be divided into three separate branches: Zetetic Analysis, Poristic Analysis and Exegetic or Rhetic Analysis. The following passage gives some idea of his ideas on this subject:

In mathematics there is a certain way of seeking the truth, a way which Plato is said first to have discovered, and which was called 'analysis' by Theon and was defined by him as 'taking the thing sought as granted and proceeding by means of what follows to a truth that is uncontested'; so, on the other hand, 'synthesis' is 'taking the thing that is granted and proceeding by means of what follows to the conclusion and comprehension of the thing sought.' And although the ancients set forth a twofold analysis, the zetetic (ζητητική) and the poristic (ποριστική), to which Theon's definition particularly refers, it is nevertheless fitting that there be established also a third kind, which may be called rhetic or exegetic (ῥητικὴ ἥ ἐξηγητική), so that there is a zetetic art by which is found the equation or proportion between the magnitude that is being sought and those that are given, a poristic art by which from the equation or proportion the truth of the theorem set up is investigated, and an exegetic art by which from the equation set up or the proportion there is produced the magnitude itself which is being sought. And thus, the whole threefold analytical art, claiming for itself this office, may be defined as the science of right finding in mathematics. Now what truly pertains to the zetetic art is established by the art of logic through syllogisms and enthymemes, the foundations of which are those very stipulations (symbola) by which equations and proportions are arrived at, which stipulations must be derived from common notions as well as from theorems that are demonstrated by the power of analysis itself. In the zetetic art, however, the form of proceeding is peculiar to the art itself, inasmuch as the zetetic art does not employ its logic on numbers—which was the tediousness of the ancient analysts— but uses its logic through a logistic which in a new way has to do with species. This logistic is much more successful and powerful than the numerical one for comparing magnitudes with one another in equations, once the law of homogeneity has been established; and hence there has been set up for that purpose a series or ladder, hallowed by custom, of magnitudes ascending or descending by their own nature from genus to genus, by which ladder the degrees and genera of magnitudes in equations may be designated and distinguished [2].

9. Napier and Logarithms

*Un grand' autore, e una invenzione grand-
issima nelle matematiche.*

Torricelli

John Napier was born in the year 1550 at Merchiston
Castle, near Edinburgh. Merchiston had been acquired by
Alexander Napier, a citizen of Edinburgh, and had belonged
to the family since before 1438. The second Napier of
Merchiston became Comptroller to King James II, and the
Scottish court twice employed him as an Ambassador.
His son was elected to Parliament, and many members of
the family had titles conferred upon them.

John Napier's father, Sir Archibald Napier, the seventh
Napier of Merchiston, increased the size of the family
estates and took an interest in public affairs. For more than
thirty years he was Master of the Mint. He was a Protestant,
and attended several General Assemblies of the Reformed
Church, but during the troubles of 1570 he hesitated between
the two factions, thus incurring the enmity both of the
supporters of Queen Mary and of the supporters of her son
James. The family castle had to endure attacks by troops
of both sides.

Sir Archibald had been married twice. John was the
eldest of the three children of the first marriage, and there
were ten children from the second.

Nothing is known about the early years of Napier's life.
He seems at first to have studied at home, but when his

mother died, in 1563, when he was thirteen years old, he
was sent to the University of St. Andrews, the oldest of the
Scottish Universities.

There were three colleges at St. Andrews: Napier was sent
to St. Salvator's College, which was under the direction of
Dr. John Rutherford, who had studied in France and had
taught at Bordeaux at the Collège de Guyenne, and in
Portugal at the University of Coimbre. Rutherford's interests were literary as well as philosophical and his works
on Aristotle are written, not in the Latin of Scholastics,
but in the more correct and elegant Latin of the Humanists.
He was, however, a violent man, and the university authorities several times found themselves compelled to call
upon him to observe moderation. He gave the young
Napier a taste for good Latin and for the study of Theology.

Napier probably did not stay long at St. Andrews, for
he did not take a degree there. It is likely that he travelled
abroad, as was the custom among young Scottish noblemen,
but all we know for certain is that by 1571 he was once more
back in Scotland.

At this time he lived not at Merchiston Castle but in
Gartness, in the parish of Drymen in Stirlingshire, where
his father owned an estate, and it is to be presumed that he
stayed there until his father's death in 1608. In 1572 he
married Elizabeth Stirling, the daughter of Sir James
Stirling, of Keir, a neighbour of the Napiers. Napier and his
wife lived at Gartness, close to the river Endrick, in a large
house built specially for them with a garden and an orchard.

There was continual unrest in Scotland owing to dis-
sension between the Catholics, who supported France,
and the Protestants, who supported England. Napier was a
staunch supporter of the Protestant cause. In 1593 he
published *A Plaine Discovery of the Whole Revelation of
St. John*, which is concerned with the Apocalypse. The work
was well received and it was reprinted in London in 1594 and

in 1611, as well as in Edinburgh (where the first edition had appeared) in 1611 and 1645. A French translation of it was published in La Rochelle in 1601 and had been reprinted nine times by 1607. Three editions of a Flemish translation were printed between 1600 and 1607, and four German editions appeared between 1611 and 1627.

Napier was reputed by some to possess magical powers. He owned a black cock which is said to have informed him of his servants' most secret thoughts. Once, when something had been stolen, he dusted this cock over with soot and put it in a darkened room. Each servant then had to go into the room and stroke the cock's back. Napier had said that the cock would crow to tell him which servant was guilty. No sound was heard, but one servant had clean hands!

Another story was told of how when a neighbour's pigeons ate seed from his fields Napier threatened to seize the birds. 'Do so if you can', replied the neighbour. Next morning, the pigeons, struck down by a powerful charm, lay motionless on the ground and Napier's servants merely had to pick them up.

In 1594, his reputation for being endowed with supernatural powers led Napier to make the acquaintance of the outlawed Robert Logan of Restalrig. Logan owned Fast Castle, a forbidding fortress perched on a steep rock whose foot was washed by the sea. James VI once said that the man who had had it built deserved to be stabbed to the heart.

There was a tradition that fabulous riches were hidden in the depths of Fast Castle and in his search for the treasure Logan turned to Napier. Their contract, which Napier drew up in his own hand, is still in existence. By its terms, Napier was, with God's help, to do all he could to discover the treasure and was to receive a third of it as recompense.

If nothing was found, he left it to Logan's discretion to pay him as he pleased for his pains.

Another of Napier's manuscripts is a list of engines of war which 'by the Grace of God and with the help of skilful workmen' he hoped to construct 'to defend this Island'. These machines included two types of burning glass, a cannon capable of killing 40 000 Turks without risking the life of a single Christian, an assault vehicle and a submarine.

Napier died on 4 April 1617, probably from gout.

His fundamental contribution to mathematics was the invention of logarithms.

We shall, however, begin by discussing a more elementary but rather interesting work, his *Rabdologia*. The basic idea is very simple. It is rather like the process of multiplication 'by parallelograms' which we shall discuss in Vol. 2, Chapter 1. The Parisian traveller Jean Chardin described a technique very similar to that of Napier's rods, which he had seen in use in Persia and in India in 1686. A description of Napier's rods is to be found in the *New Epitome of Arithmetic* (*Nouveau Epitome d'Arithmétique*) (Liège 1616) written by the military engineer, Jean Gallé.

Napier's rods form a sort of multiplication table, written out on rods with a square section. They enable partial products to be written out very rapidly, and the process of multiplication is essentially reduced to one of addition.

The title page of Napier's work reads:

Rabdologia, seu Numerationis per Virgulas Libri Duo: Cum Appendice de expeditissimo Multiplicationis Promptuario. Quibis accessit et Arithmeticae Localis Liber Unus. Authore et Inventore Ioanne Nepero, Barone Merchistonii etc. Scoto. Edinburgi, Excudebat Andreas Hart, 1617.

The appendix describes an improved system in which the rods are replaced by perforated plates; this made it possible to multiply very large numbers. The book called *Arithmetica Localis* describes a very strange use of binary numbers for making calculations by using counters on a chess board. A second edition of Napier's book was printed in Leyden in 1626, an Italian translation, by Marco Locatello, came out in Verona in 1623, and a German one, by Benjamin Ursinus, appeared in Berlin in the same year.

It is unfortunately impossible for us to go into the technical details of the invention of logarithms, to which Napier devoted twenty years of determined effort.* He finally published his results in 1614, in a work entitled *Mirifici Logarithmorum Canonis Descriptio, Ejusque usus in utraque Trigonometria; ut etiam in omni Logistica Mathematica, Amplissimi, Facillimi, et expeditissimi explicatio. Authore ac Inventore, Ioanne Nepero, Barone Merchistonii, etc. Scot. Edinburgi, Ex officina Andreae Hart, Bibliopolae* CIƆ, *DC, XIV.*

The author was justly proud of his work, and in the dedicatory letter addressed to the Prince of Wales he writes: 'As you receive the benefit of this little work, tender praise and thanks to God, the Almighty Author and Giver of all good things.'

The work was reprinted in 1619 after the author's death. This second edition, printed in Edinburgh, also included the *Mirifici Logarithmorum Canonis Constructio*, which complements the *Descriptio* by explaining how to compute tables of logarithms.

In the same year, 1619, the two treatises were again published together in Lyon. The *Descriptio* had been translated into English by Edward Wright and published in London in 1616, after the translator's death, under the title *A Description of the Admirable, Table of Logarithms, With a declaration of the most Plentiful, Easy and speedy use thereof*

* Laplace's remarks which we quote on page 308 give some further information about logarithms.

in both kindes of Trigonometrie, as also in all Mathematical calculations. Invented and Published in Latin by That Honourable L. John Napier Baron of Marchiston, and translated into English by the late learned and famous Mathematician Edward Wright, With an Addition of an Instrumentall Table to finde the part proportionall, invented by the Translator, and described in the end of the Booke by Henry Briggs, Geometry-reader at Gresham House in London. All perused and approved by the Author, and published, since the death of the Translator. London, printed by Nicholas Okes, 1616.

Edward Wright (1560–1615), a mathematician employed by the Right Worshipful Company of Merchants of London trading to the East Indies, was known as a traveller and a cartographer: he gave the first satisfactory explanation of Mercator's projection,* which shows loxodromic lines as straight ones. Strangely enough, his work on the subject, published in 1599, contained, unknown to Wright, an anticipation of the idea of logarithms.

Henry Briggs (1556–1630), mentioned in the title of the English translation of the *Descriptio*, was, like Wright, exceedingly interested in Napier's work. He entered into correspondence with Wright on the subject and they agreed that certain modifications should be made to Napier's system. The result was that they drew up tables of decimal logarithms like those in use today.

Rarely has a discovery spread as rapidly as this one did. More than twenty works on logarithms were published between 1614 and 1631.

In all justice, it should be pointed out that Jobst Bürgi (1552–1632) had invented logarithms independently of

---

* Invented by the Flemish cartographer Gerhardus Mercator (Gerhard Kremer) (1512–1594), who is not to be confused with the mathematician, who came from Holstein, Nicolaus Mercator (Kauffman) (1620–1687).

Napier.* Bürgi's tables were not published until 1620, but they had been drawn up before 1610.

The first French treatise on logarithms appeared in 1625. It was written by an Englishman, Edmund Wingate,† and was dedicated to 'The Most High and Most Powerful Prince, Monsieur, Gaston de France, only Brother to the King, Duke of Anjou, etc. . . . '

In a short preface, a summary of the *Arithmetica Logarithmica* of Briggs, of 1624, logarithms are defined and explained. Napier is not mentioned, but Briggs is given the credit for having invented logarithms with base 10.

In Paris in the following year Denis Henrion published his *Treatise on Logarithms* (*Traicté des Logarithmes*).

In 1624, an Englishman called Gunter‡ had the idea of having a logarithmic scale engraved on copper. This primitive version of the slide rule, using only one scale, immediately found wide acceptance, in particular by the British Navy. It was used in conjunction with one, or possibly two, pairs of dividers.

In 1626, Wingate produced an improved version of Gunter's rule, using two scales which could slide against one another.

In 1662, an Englishman called Seth Partridge gave a description of a rule he had made in 1654, employing a small scale which slid between two fixed ones. This was essentially the modern slide rule.

---

* Jobst Bürgi, a Swiss, was born in Lichtenstein on 23 February, 1552. He was a clock-maker and an inventor. Jean III Bernoulli wrote of him in 1768: "The most curious piece in the room containing clocks is, quite certainly, an automatic astronomical device made by the ingenious Juste Byrgius for William IV. It is equally admirable for its invention and for its execution. It is really amazing and even several pages would not suffice to describe it. A detailed study of it would be a complete course in Ptolemaic astronomy and to understand its working one would have to have an uncommonly good grasp of the principles of clock-making."

† Edmund Wingate was born in Yorkshire in 1596. He died in 1657.

‡ Edmund Gunter (1581–1626).

The instrument rapidly became popular in England, but was accepted much more slowly in France. Prony,* who in 1793 had been responsible for the calculation of the new trigonometric tables using centesimal divisions, wrote in a popular work published in 1833 that:

'From this point of view, the English have a sort of superiority over the French. National honour and commercial interest demand that it be abolished. In London, in many shops, and even in some of the lowest class, I saw tables of logarithms on the counter. Foremen and workmen are provided with equivalents of these tables engraved on sliding rulers, and with these they can easily and very quickly perform calculations which could not be performed with the same expedition by means of pen and paper. Some French scholars, among whom I must mention M. Jomard, a member of the Institut Royal de France, have made efforts to introduce these sliding rules into our own workshops, but they have not been as successful as one would have hoped.'

The slide rule took the place of another mathematical instrument: the sector, which was much used in the seventeenth century. Galileo describes the instrument in a work published in Padua in 1606 and entitled *Le operazione del Compasso geometrico e militare*. He claimed to have invented the sector in about 1598. though in fact such instruments had been known for some thirty years. Galileo made several sectors from 1599 onwards.

In Padua in 1607 a certain Baldassare Capra published a short pamphlet entitled *Usus et fabrica circini cujusdam proportionis*, which is scarcely more than a Latin adaptation of Galileo's work. The author claims to have invented the instrument, says that he has been making examples of it for the last ten years and accuses Galileo of plagiarism.

---

* Baron Riche de Prony (1755–1839), a French mathematician and engineer.

During the years 1610–15, Alleaume, the King's engineer (the same Alleaume who had inherited Vieta's manuscripts) made sectors in Paris. They were less sophisticated than those of Galileo.

A sector consists of two flat rulers of equal length which can lie parallel to one another when the sector is closed, and are joined together at one end by a hinge. When the sector is completely open, the two rulers lie exactly in a straight line. From the axis of the hinge, straight lines fan out along the two rulers. These lines carry various scales, designed for different purposes, and the instrument is symmetrical. There is, for instance, a scale divided into equal parts, along which graduation is in arithmetic progression, and a scale for areas where the distances from the axis are proportional to the square roots of the numbers written on the scale, and a scale for volumes where the distances are proportional to the cube roots of the numbers etc. The instrument was used in conjunction with dividers (see plate XI).

This plate is in fact an illustration for the following problem, taken from Manesson Mallet's *Practical Geometry* (*Gèométrie Practique*, 1702):

To divide the circumference *ABCDE* into five equal parts.

First, use dividers to find the length of its semidiameter *FB*. Then, open the sector, and using the scales that are equally divided place the points of the dividers between the two sixth marks of the lines indicating polygons. Leave the sector open in this position.

Then, looking along these same lines indicating polygons, we use dividers to find the distance between the two fourth divisions (if we want to divide the circumference into four equal parts) or the opening between the two fifth divisions (if, as in the present problem, we want to divide the circumference into five parts). The distance found with the dividers will cut the given circumference into five precisely equal parts, at the points, *A*, *B*, *C*, *D* and *E*.

PLATE XI. The use of a sector. From Manesson Mallet, *Géométrie Practique* (1702).

10. The Golden Age

Le XVIIe siècle, celui de tous qui fait le plus
*d'honneur à l'esprit humain**
Laplace.

In this chapter we shall discuss some French mathematicians of the seventeenth century. By fortunate historical chance we shall thereby include some of the greatest scholars of all time.

10.1. Pierre Fermat

Mi par di veder un gran lume.†
Fermat

From the *Journal des Sçavans*, of Monday, 9 February 1665.

We were very sorry to learn today of the death of M. de Fermat, Councillor of the Parliament of Toulouse. He was one of the finest minds of the century, and his genius was so universal and so wide in its extent that, if scholars did not all agree in acknowledging his extraordinary merit, it would be difficult to believe all that should truly be said of him, if we are to give him the praise he deserves.

He carried on a personal correspondence with Descartes, Torricelli, Pascal, Frenicle, Roberval, Huygens and others, as well as with the majority of the great geometers of England and Italy, but his closest friend was M. de Carcavi, whom he met when they were colleagues at the Parliament of Toulouse. Carcavi, formerly Fermat's companion in study, now finds himself the heir of all his friend's papers.

* The seventeenth century, which was the time of the greatest achievements of the human spirit.

† I think I see a great light.

But since the purpose of this journal is mainly to make known through their works those who have become famous in the republic of letters, we shall content ourselves with giving a catalogue of the writings of this great man, and shall leave it to others to write fuller and more worthy tributes.

He excelled in all branches of mathematics, but particularly in number theory and in geometry. He invented a method of squaring parabolas of any degree.

A method of maxima and minima which can be used not only for plane and for three-dimensional problems but also enables one to find the tangents to curves and the centres of gravity of solids, and to solve numerical problems.

An introduction to loci, plane and three dimensional, which is an analytical treatise on the solution of problems in two and three dimensions. This was written before M. Descartes had published anything on the subject.

A treatise *de contactibus sphaericis* in which Fermat proved in three dimensions theorems which M. Viète, the Referendary, had only proved for plane figures.

Another treatise in which he states and proves the propositions of Apollonius Pergaeus's two books on plane loci.

And a general method for finding the length of curved lines, etc.

Moreover, he was very well acquainted with the Classics, and was often consulted in case of difficulty. He explained any number of obscure passages in ancient writings. Some of his notes on Athenaeus have recently been printed, and the translator of Benedetto Castelli's work on measuring running water* inserted into the work a brilliant note of Fermat's on a letter of Synesius. The letter itself was so obscure that Père Petau, who wrote a commentary on Synesius, admitted that he could not understand it. Fermat also wrote many notes on Theon of Smyrna and other ancient authors. Most of these notes are to be found scattered among his letters because he did not often write on this sort of subject except in answer to requests from his friends.

His mathematical writings and his research into the Classics did not interfere with M. de Fermat's devotion to his profession. indeed he carried out his duties so well that he passed for one of the best lawyers of his time.

But what is even more surprising is that despite having the strength of character inseparable from the rare qualities of mind which we have mentioned, he was also endowed with a sensibility which enabled him to write Latin, French and Spanish verse with as much elegance as if he had lived in the time of Augustus, or had passed his life at the Court of France or at the Court in Madrid.

We shall present a more detailed discussion of this great man's work when we have collected together his published writings and have obtained his son's permission to publish the works which are still in manuscript.

---

* Saporta.

This tribute makes it unnecessary for us to go into many details about Fermat's life.

Pierre Fermat was born at Beaumont-de-Lomagne in August 1601. He became an Advocate, and then, from 1631 onwards, a Magistrate in Toulouse. He was married, and had five children. He was a good magistrate, and when he died at Castres on 12 June 1665, in the course of an official journey he would have been no more than an unknown honest man had he not had a genius for mathematics.

Fermat, the most important of Vieta's pupils, adopted Vieta's mathematical notation. Together with Descartes he invented analytical geometry, and together with Blaise Pascal the calculus of probabilities. Like Cavalieri and Roberval, he was one of the masters of that primitive form of integral calculus known as the *method of indivisibles*, but his work also led towards the differential calculus and he was famous for his fundamentally important work in number theory or higher arithmetic.

There is a proposition in number theory which is known as *Fermat's Lesser Theorem*. It states that:

If a is an arbitrary integer and p is a prime number then $a^{p-1} - 1$ is divisible by p.

Fermat stated this theorem in a letter to Frenicle dated 18 October 1640. The proof is elementary.

Fermat's Greater Theorem states that if x, y, z and n are all integers the relation $x^n + y^n = z^n$ is possible only if n is less than or equal to 2.

This proposition was never published by its author, who merely wrote it in the margin of his copy of Diophantus. He had however mentioned it in his correspondence, though only for the special cases $n = 3$ and $n = 4$.

This theorem gave rise to a considerable literature, but it has still not been proved for the general case.

In Volume 2, Chapter 4 we shall quote the passage where Fermat proves that it is impossible to find a right angled triangle whose sides are integers and whose area is a perfect square. This is equivalent to proving his Great Theorem for the case $n = 4$.

Fermat made an interesting contribution to the development of differential calculus. The following passage* is taken from a work on finding maxima and minima. The reader must bear in mind that the algebraic notation used is not the author's own.

METHOD OF MAXIMUM AND MINIMUM

Whilst studying Vieta's method of *syncrisis* and *anastrophe* ('combination' and 'negation') and exploring more carefully their use for determining the structure of correlated equations, I had the idea of using it to work out a procedure for finding a maximum or a minimum, thereby overcoming all the difficulties encountered with limiting conditions, difficulties which have been very troublesome to ancient as well as to modern geometers.

Maxima and minima are indeed unique and single points, as Pappus says, and as was already known to the Ancients, although Commandinus admits that he does not know what Pappus means by the term $\mu o \nu a \chi \acute{o} \varsigma$ (unique).†
It follows from this that on each side of the limiting point we may take an ambiguous equation and that the two ambiguous equations taken in this way are correlated, equal and similar.

Let us suppose, for example, that it is required to *divide the straight line b in such a manner that the product of the segments shall be a maximum*. The point which satisfies this requirement is clearly the mid-point of the given straight line, and the maximum product is equal to $b^2/4$; no other way of dividing the straight line will give a product equal to $b^2/4$.

---

* Originally in Latin. We have translated from Paul Tannery's French version. J.V.F.

† In another memoir Fermat writes, in French: It is here that Pappus calls minimam proportionem $\mu o \nu a \chi \grave{o} \nu \ \kappa a \grave{\iota} \ \acute{\epsilon} \lambda \acute{a} \chi \iota \sigma \tau o \nu$ minimam et singularem. The reason for this is if the problem is posed in terms of given quantities there are always two possible solutions. For the smallest or the largest value there is only one solution to the problem. This is why Pappus uses the words *minimam et singularem* to mean unique, the least proportion of all those which could be considered. In this passage Commandinus is unsure 'what Pappus means by $\mu o \nu a \chi \acute{o} \varsigma$'; I think that perhaps seeing the word 'monk' together with the word 'minimus', he thought about Père Mersenne's order [the Minims. J.V.F.].

But if we are required to divide the same straight line b in such a way that the product of the segments shall be equal to an area z (this area must be presumed to be less than $b^2/4$), there will be two points which satisfy the requirement, and they will be situated one on either side of the point corresponding to the maximum.

Let a be one of the segments of the line b, then we have $ba - a^2 =$ area z, which is an ambiguous equation since each of its two roots may be taken as the straight line a. Let the *correlated* equation be $be - e^2 =$ area z. Let us compare these two equations by Vieta's method:

$$ba - be = a^2 - e^2.$$

Dividing both sides of the equation by $a - e$, we obtain $b = a + e$, where the lengths a and e are to be unequal.

If instead of some area z we take another, larger, area, also less than $b^2/4$, the new straight lines a and e will differ from each other less than the previous ones, the dividing points coming closer to the point corresponding to the maximum product. The more the product of the segments increases the more the difference between a and e will decrease, until it vanishes completely for the division corresponding to the maximum product; in this case, there is only a single solution, since the two quantities a and e become equal.

Now, when we applied Vieta's method to the two correlated equations above, it gave us the equation $b = a + e$; so if $e = a$ (which is always true for the point corresponding to the maximum or the minimum), we shall have, in the case considered, $b = 2a$, which is to say that if we take the mid-point of the straight line b the product of the segments will be a maximum.

. . . . .

However, since in practice division by a binomial is generally a complicated and laborious process, it is preferable, when comparing correlated equations to write them as far as possible in terms of the difference of the roots, so that we only require to divide through by that difference.

Suppose we are required to find the maximum of $b^2a - a^3$. According to the rules of the method described above we should take as our correlated equation the equation $b^2e - e^3$. But since e, like a, is an unknown quantity there is nothing to prevent us from writing it as $a + e$; we thus obtain:

$$b^2a + b^2e - a^3 - e^3 - 3a^2e - 3e^2a = b^2a - a^3.$$

It is clear that if we combine similar terms all the remaining terms will contain the unknown e, since the terms which only contain a are the same on both sides of the equation. So we have:

$$b^2e = e^3 + 3a^2e + 3ae^2,$$

and, dividing through by e,

$$b^2 = e^2 + 3a^2 + 3ae,$$

which gives us the form of the two correlated equations.

To find the maximum, we must set the roots of the two equations equal to one another, in accordance with the rules of the original method, from which our new procedure derives both its logical structure and its mode of operation.

So we must set a equal to $a + e$, which makes $e = 0$. But, from the form we found for the correlated equations,

$$b^2 = e^2 + 3a^2 + 3ae;$$

so all the terms in e that occur in this equation must be removed, since they reduce to zero; we are left with $b^2 = 3a^2$, an equation which will give the required maximum for the product concerned.

Fermat used his method of finding maxima and minima, the prototype of the calculation of a differential coefficient, in order to find tangents to curves. For this he appealed to analytical geometry, to that *Introduction to loci, plane and three dimensional, an analytical treatise on the solution of problems in two and three dimensions* mentioned by the Journal des Sçavans. He adopted the Greek definition of a tangent (see above p. 80) and was clearly thoroughly familiar with Archimedes's *On the Spiral*, which makes such skilful use of this definition. In order to follow Fermat's argument, which most of his contempories were unable to do (on this occasion even Descartes failed to recognize Fermat's achievement) we shall consider an example.

We are required to draw a tangent at a particular point on the curve $ax^2 = y^3$ (using Cartesian notation). Let M be the point on the curve where we are required to construct the tangent TM (Fig. 10.1). Let O be the origin, and $TOPp$ the diameter of the curve (along the x axis). The ordinates (lines such as MP) are parallel to each other, though not necessarily perpendicular to the diameter.

At a point M on the curve where $OP = x$ and $PM = y$ we have $y^3 - ax^2 = 0$.

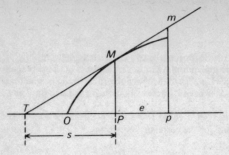

Fig. 10.1

At a point m, on the tangent TM, y^3 is greater than ax^2, since the tangent lies entirely above the curve. Also, if m moves along this tangent, the difference in question $y^3 - ax^2$ goes through a minimum, zero, value at M.

This properly allows us to construct the tangent. Let TP be the subtangent, whose length s we require to find. Let $Pp = e$. Triangles TMP and Tmp are similar to one another,

$$\therefore \quad \frac{mp}{MP} = \frac{Tp}{TP}$$

i.e.

$$\frac{mp}{y} = \frac{s + e}{s}$$

so $mp = y(1 + e/s)$.

For a minimum of $(mp)^3 - a(Op)^2$, which is reached at the point M, we require that the coefficient of e shall be zero in the expression

$$y^3 \left(1 + \frac{e}{s}\right)^3 - a(x + e)^2$$

i.e. we require that

$$3\frac{y^3}{s} - 2ax = 0$$

i.e.

$$s = \frac{3y^3}{2ax}.$$

Now $y^3 = ax^2$,

$$\therefore \quad s = \tfrac{3}{2}x.$$

Thus $TP = \tfrac{3}{2}OP$ and $TO = \tfrac{1}{2}OP$.

We should note that this method gives us a necessary condition for the line TM to be a tangent to the curve, that is, it proves not the existence of a tangent at M but its uniqueness.

We mentioned above that Descartes did not completely understand Fermat's method, and this was, in fact, partly Fermat's fault. Descartes had also worked in analytical geometry and was the most perceptive of Fermat's contemporaries. In his *Geometry* he had described an analytical method of finding normals to curves and he had used an analogous method to check Fermat's result. He considered the line TM, which cuts the curve in M and m (Fig. 10.2).

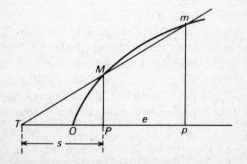

Fig. 10.2

Thus, using the same lettering as before, we again have $y^3 - ax^2 = 0$. Also, $y^3(1 + e/s)^3 - a(x + e)^2 = 0$,

$$\therefore e\left[\frac{3y^3}{s} - 2ax\right] + e^2\left[\frac{3y^3}{s^2} - a\right] + e^3\frac{y^3}{s^2} = 0.$$

Dividing throughout by e we obtain

$$\frac{3y^3}{s} - 2ax + e\left[\frac{3y^3}{s^2} - a\right] + e^2\frac{y^3}{s^3} = 0.$$

The attitude Descartes adopts to tangents is modern, different from that of the Ancient Greeks: for him a tangent cuts a curve in two coincident points, at the point of contact. Thus we can find the tangent by substituting $e = 0$ in the above equation. This gives us $3y^3/s - 2ax = 0$, as Fermat required. In fact, Descartes is not doing his rival an injustice. He is simply applying Fermat's method of finding maxima as a direct method of finding tangents. This shows that the ideas of the differential calculus were already beginning to develop as early as 1638.

10.2. Roberval

Gilles Personne or Personnier de Roberval was born in the village of Roberval, near Sentis on 10 August 1602. In his youth he travelled a great deal: 'studying and teaching', he tells us. He settled in Paris in 1628 and in 1638 became a philosophy teacher at the Collège de Maître Gervais in the 'rue de Foing by the church of the Mathurins'.* He never married and remained at the College until his death in 1675. He occupied the Chair founded by Ramus at the Collège Royal from 1634 until 1675 and he was a member of the Académie des Sciences from the time of its foundation in 1669.

---

* The Church of the Mathurins was in the rue Saint Jacques, next to the Hôtel de Cluny. The Collège de Maître Gervais was founded in 1370 by Maître Gervais Chrétien, chief physician to King Charles V.

Roberval was a very distinguished scholar. He was also a rather stubborn character, and a personal enemy of Descartes. He published little important work, but he had many new ideas, and since several of these ideas occurred to other people at much the same time there were inevitably quarrels about priority.

From his work in mechanics, Roberval gave his name to a type of balance widely used in commerce. As for geometry: he invented a kinematic method of finding tangents which put to skilful use. In modern vectorial terms the method might be said to use the fact that the absolute velocity is the sum of the relative velocity and the common velocity. This method enabled Roberval to construct tangents to the cycloid, to the conics (making use of the bifocal definition), to Archimedes's spiral, to Dinostratus's quadratrix and to Nicomedes's conchoid. Using Vieta's methods to investigate the properties of this last curve Roberval showed that it had points of inflexion.* He also constructed the tangents to Pascal's limaçon curves.

At this time, Archimedes's letter to Eratosthenes (*The Method*) had not yet been rediscovered (it was not even known to exist) and nothing was known about Archimedes's heuristic and analytical methods. Several geometers calculated areas and volumes, and determined centres of gravity, and all the analytical methods they developed for the purpose, under the general name of the *Method of Indivisibles*, had in fact been employed by Archimedes.

An Italian priest, Bonaventura Cavalieri (1598–1647) was the first to publish general results and to explain his technique. His *Geometria indivisibilibus continuorum nova quadam ratione promota* (published in Bologna in 1635) excited universal interest. Cavalieri's work is not entirely

---

* He described this discovery to Fermat, who pointed out that points of inflexion on a curve could be found by finding a maximum or minimum in the gradient of the tangent.

rigorous, but this is not an important failing because once the results have been obtained the well-established Greek methods of exhaustion enable rigorous proofs to be given. A rather more serious flaw is Cavalieri's refusal to use algebra in geometrical work. Torricelli, a pupil of Cavalieri, followed his master's example, and this rejection of algebra not only prevented them from noticing certain things but also made some of their arguments unnecessarily clumsy.

Their French contemporaries, Fermat, Descartes and Roberval, did not have the same scruples, and their progess was often more rapid than that of the Italians.

The two groups did not have any contact with each other for a long time (at least until after 1640). For example, Cavalieri wrote to Torricelli on 22 September 1643:

> Roberval's name was unknown to me, but you say that he is a famous geometer. I am very much surprised that neither Father Niceron nor Giovanni Antonio Rocca, who wrote me a long letter giving me a list of all the most important mathematicians in Paris, ever mentioned him at all. All the same, this does not affect my esteem for him, since you consider him to be an eminent man. Indeed, it cannot but be true, since he has proved the results you mention, and in particular those marvellous propositions of yours. Moreover, I remember that Galileo, of blessed memory, told me once that he enquired of some important Parisian gentlemen for news of François Viète and no-one recognized the name. But when he remembered that Viète was a Councillor and the King's Secretary, then they remembered him. They had never heard of his being a mathematician.

We shall now turn to one of Roberval's discoveries, which he described to Torricelli in 1646.

Let AMB be a convex curve with diameter Ab (Fig. 10.3). The tangent at M is the line MT, which meets the diameter Ab at the point T. The line MN is parallel to Ab, and NT is the perpendicular to Ab at T. The locus* of N is the curve ANC. If m is a point infinitely close to M on the first curve we may assume that mM is on the tangent MT 'since small

---

* The expression 'lieu géométrique' (geometrical locus) was invented by Roberval.

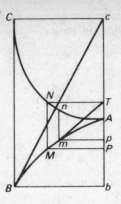

Fig. 10.3

segments of curved lines become straight if they are infinitely divided.' From this we deduce that the areas $MPpm$ and $MNnm$ (see figure) are equivalent infinitely small areas.* Roberval deduces that the areas $AMBb$ and $AMBCNA$ are equal. Now if the original curve is a parabola $y^m = a^{m-p}x^p$, Fermat's method explained above shows that

$$AT = AP \times \frac{m-p}{p}.$$

The curve ANC is therefore related to the original curve by an orthogonal transformation with the ratio $(m-p)/p$. Taking area $AMBb = k \times Ab \times Bb$ we therefore have area $AMBCNA = k \times Ab \times Bb$ and area

$$ANCc = \frac{m-p}{p}k \times Ab \times Bb.$$

---

* Euclid, *Elements*, Book I, Proportion 43.

But the sum of these three areas is the rectangle $bBCc$ and $bc = m/p \times Ab$. We thus have

$$\frac{m}{p} = k\left(\frac{m-p}{p} + 2\right)$$

$$= k\left(\frac{m+p}{p}\right).$$

$$\therefore \quad k = \frac{m}{m+p}.$$

In particular, for the ordinary parabola, where $m = 2$ and $p = 1$, $k = \frac{1}{3}$. This is the result obtained by Archimedes. Thus the work done by Fermat and Roberval, who had great respect for each other, could have led to an early realization of the connection between integral and differential calculus (though Leibniz's terms are anachronisms in this context). This connection was not in fact understood until the end of the century.

Torricelli, who died in 1647 at the age of thirty-nine, knew about the work of the two Frenchmen, but did not realize its true importance. His friend Ricci, however, was far more perceptive, and through him Newton's teacher, Isaac Barrow, was to become aware of the new methods.

10.3. Descartes

Descartes et Newton, ces deux
*législateurs de l'art de penser**
            D'Alembert

René Descartes, who was born at La Haye, in Touraine, on 31 March 1596, came from a family of magistrates. From Easter 1604 until August 1614 he studied at the Jesuit college of La Flèche, where the teaching of mathematics was inspired by the work of Clavius and the teaching

---

* Descartes and Newton, those two lawgivers of the art of thinking.

of algebra by the work of Stifel, through that of Clavius. At this time mathematics, taught as part of the Philosophy course, was only studied by a very small number of pupils. Descartes's teacher at La Flèche, Maître Jean François, was a very young man who was still studying theology. He is mentioned later as a Jesuit Father, in Paris at the Collège de Clermont, now the Lycée Louis-le-Grand, from 1620 to 1624.

At La Flèche an able student such as Descartes could obtain an education roughly equivalent to that given in secondary schools to-day.

Descartes stayed for a little while in Paris, then studied law in Poitiers from 1615–16, and towards the end of 1617 went to Breda to attend the Military Academy. There he met the scholarly Isaac Beeckman, and came into contact with mathematicians of the Flemish school, including Stevin, and the great French algebraist, the engineer Albert Girard. This school still employed the exponential notation used in the fifteenth century by Nicolas Chuquet and adopted by Bombelli in Italy.

Descartes left Breda in April, 1619. He travelled through Denmark and Germany, where he met the mathematician Johannes Faulhaber (1580–1635), and on 8 November 1620, was present at the Battle of Prague. He definitively abandoned a military career in 1621, and after further travels in France, Switzerland and Italy, settled in the Netherlands, in 1628, though he did pay some brief visits to France. In 1649 he accepted an invitation to visit Queen Christina of Sweden. He died in Stockholm on 11 February 1650.

Like Pascal (and unlike Fermat and Roberval) Descartes was rich enough not to depend for his livelihood on following a profession. We shall now give a more detailed account of his life and of the evolution of his thought, describing the development of the bold ideas which were to lead him to the

startling discoveries he himself greeted with such enthusiasm [1].

The time that Descartes spent at the college of la Flèche,
from 1604 until 1614, marked the beginning of his interest
in the Sciences and of his desire to free them from their
purely subsidiary rôle.

> I felt a great esteem for eloquence and I loved poetry deeply, but I
> thought that both of them were natural gifts rather than the fruit of study....
> I particularly enjoyed mathematics because the arguments were clear and
> convincing, but I was not aware of its true scope and, thinking it could be
> applied only to engineering, I was surprised that on such firm and solid
> foundations nothing more elevated had been built.

This is typical of Descartes's thought, not only in the
attitude it shows to mathematics, but also in the way it
indicates his tendency to feel that the clarity of the underlying principles of mathematics made them a suitable basis
for a philosophy in which everything would be reducible to
numbers and to axioms. It took Descartes long years of
thought and study to work out such a system but as early
as the beginning of 1619 he said in a letter to Beeckman
that, although he realized the difficulty of carrying out his
design, he had glimpsed the hope of 'producing an entirely
new science in which all the problems involving any type of
continuous or discrete quantity could be solved'. And
later: 'The work to be done is, immense, and it cannot be
done by one man. It shows incredible ambition but I have
seen I know not what light shining through the chaos of
this science of mine and I hope that with its help I shall be
able to clear up even the most difficult obscurities.'*

The development of Descartes's system is marked by
two curiously related dates: 10 November 1619 and
10 November 1620. Descartes's first biographer, the Abbé

---

* 'Infinitus' quidem opus est, nec unius. Incredible quam ambitiosum,
sed nescio quid luminis per obscurum hujus scientae chaos aspexi cujus
auxilio densissimas quasque tenebras discuti posse existimo.'

Adrien Baillet, says 'The search for truth threw his spirit into violent agitation, which increased more and more from the continual strain, for he would not allow himself to be diverted from his work either by exercise or by company. He so tired his spirit that his brain became inflamed and he fell into a kind of manic enthusiasm which so affected his mind, weakened as it was, that he was soon in a state such that he could be influenced by dreams and visions.'

On 10 November 1619 Descartes was seized by strange feelings of illumination and exaltation at a marvellous revelation—in his discourse *Olympica* he even seems to call it a miraculous revelation: 'cum plenus forem Enthousiasmo et mirabilis scientiae fundamenta reperirem'.

That night Descartes was greatly moved by three dreams which he himself explained, recorded and described. This enabled Baillet to give a detailed account of them. Here we shall merely mention that in the third dream Descartes was offered a book of poetry in which he came upon the line:

Quod vitae sectabor iter?*

while at the same moment an unknown man presented him with another piece of verse beginning with 'Est et non', the yes and no, the ναὶ καὶ οὔ of Pythagoras, which must be taken as symbolic of truth and falsehood in human knowledge.

Exactly a year later, on 10 November 1620, Descartes had a new revelation, which he greeted with the same enthusiasm. There was, however, a small but no doubt important difference. Descartes no longer speaks of a 'science', but of an 'invention': 'XI Novembris 1620, coepi intelligere fundamentum Inventi mirabilis'.

The secret of this invention was kept for a long time, but, according to Gustave Cohen, the means of arriving at the

---

* What road shall I follow in life?

Method, the unified science, conceived on the preceding 10 November, must have been application.

For his part, Charles Adam, commenting on Descartes's letter to Beeckman about the first revelation of 10 November 1619, saw several good reasons for the scholar's enthusiasm: 'The Pythagoreans were haunted by the idea of a mathematical system in the Universe, to be found by studying ratios and proportions between figures, numbers, heavenly bodies or sounds, Descartes has now given us such a mathematical system and it is a sufficient explanation for his enthusiasm.'* And again: 'All quantities which are related to one another numerically can be expressed in lines' which, as Paul Tannery has pointed out, might tempt us to consider that Descartes's great achievement was not to have applied algebra to geometry, but to have applied geometry to algebra.

Descartes finally aimed at this unification of all science through the second revelation which came to him on 10 November 1620.

We cannot give a full description of Descartes's numerous journeys or of the progress of his thoughts; for instance, on 8 October 1628, he was in Dordrecht, where he met Beeckman whom he had not seen for nine years. Beeckman records in his diary that during this visit Descartes said 'that as for Arithmetic and Geometry he was satisfied, that is, in those branches of Mathematics he had made as much progress as a man could in nine years'. We must suppose that in the excitement of a passionate and animated conversation a remark like this might, for once, have escaped the lips of a usually modest man. Beeckman clearly continued to have a high opinion of Descartes, since he wrote in the same diary 'I think that the reason there are so few learned men here is that those who are gifted are eager to

---

* 1619 is the date of Kepler's *Harmonicae Mundi*. J.V.F.

describe their discoveries the moment they have made them, and they are not content with merely publishing their own original work but describe the development of the subject from its origin and then add their further contribution at the end, though this useless and repetitive labour crushes spirits otherwise perfectly capable of making many further advances.

He, on the other hand, has not written anything yet, but by keeping his thoughts to himself up to his thirty-third year he seems to have succeeded better than the others in finding what he sought. I say this to encourage people to imitate his example rather than the crowd of scribblers.'

Descartes gradually became acquainted with many eminent mathematicians such as Fermat, Roberval, Pascal, Father Mersenne and Huygens, and, in the manner of the time, the correspondence often involved very animated or at times even exceedingly impolite discussions. For instance, Descartes quarrelled with his very old friend Beeckman, and also wrote the following judgment of Fermat:

M. Fermat is a Gascon;* I am not. It is true that he has done much good work and that he is a very intelligent man. But, for my part, I have always tried to consider things very generally so as to be able to deduce from them rules which could be used elsewhere.

The year 1637 finally saw the publication of his great work: the *Discourse on Method* (*Discours de la Méthode*). Descartes's first choice of title had been 'Project for a universal science which could raise our nature to its highest degree of perfection' ('Projet d'une Science universelle qui puisse élever nostre nature à son plus haut degré de perfection. . . .')

Descartes says of the title he finally adopted:

---

* Gascons were famous for boasting. J.V.F.

I do not call it *Treatise on Method*, but *Discourse on Method*, meaning something like Preface or Description of the Method, to show you that I have no design to teach it, but only to discuss it.

'Descartes's publisher was to grow rich, but the author did not ask for royalties. He asked only to be given two hundred free copies of the book, and this was the price paid for one of the purest masterpieces ever to be produced by the human spirit.' This comment is made by Gustave Cohen in the thesis from which our quotations have been taken. We are also indebted to M. Cohen for the text of the publisher's contract recently rediscovered in the Municipal Archives of Leyden, where it had previously remained unknown and unpublished until then.

Protocol 335
Laurens Vergeyl
Notary Public in Leyden

Today the 2nd December [1636] appeared before me LAURENS VERGEYL notary public and the witnesses named below M. RENÉ DES CARTES at present living in this city and Sr. JEAN LE MAIRE, a bookseller of this said city of Leyden. These two declared that they have agreed with one another that the said DES CARTES should put in the hands of the said LE MAIRE the complete manuscript of a book entitled *La méthode*, etc., also the *Dioptrique*, the *Météores* and the *Géometrie* and will join with him in arranging for it to be printed in this country and in France on condition that the said LE MAIRE will only be allowed to print two editions, the one which has already been begun in this city and another which he may print either here or in France, and that these two editions together must not come to more than 3000 copies, and when they are distributed or when the said DES CARTES is willing to buy all those which remain to the said LE MAIRE for the price for which he would otherwise have sold them to bookshops, the said DES CARTES would then own the books by the same right as if he were their publisher, and would be allowed to take them to him, LE MAIRE, or to any other bookseller he pleases, so that if after that the said LE MAIRE printed the said book, either in French or in any other language, without the consent of the said DES CARTES, he would render himself liable to the same penalties or fines to which any other person would have been subject if he had infringed the printer's privileges during the distribution of the first two editions. And in addition LE MAIRE promises to give to the said DES CARTES two hundred copies of the first edition, which is begun, and undertakes to fulfil all the conditions heretofore stated against liability in person and in his property, nothing excepted, being willing to surrender them to any justice or enquiring magistrate etc. . . .

Drawn up in Leyden in the office of me a notary in the presence of DAVID GATOU and JEAN DESPUY who are trustworthy witnesses, as it was required of me as notary.

> RENÉ DESCARTES
> JAN DU PUIS
> GATOU for the above
> JAN MAIRE.

The *Discourse on Method* was original in another important respect: it was written in French; although tradition demanded that Latin be used, as indeed it was in Gassendi's treatises, in Bacon's *Novum Organum* and the *Geometry* of Clavius.

Descartes explains himself as follows:

And if I write in French, the language of my country, rather than in Latin, the language of my tutors, I do so in the hope that those who use only their natural reason are better equipped to judge my ideas than those who only believe in venerable books, and the only people I wish to have as my judges are those who have commonsense as well as learning, who, I am sure, will not prefer Latin so much as to refuse to follow my reasoning merely because I explain myself in the vernacular.

We shall now return to mathematics and discuss Descartes's work as tutor to Princess Elizabeth, the daughter of Frederick of Bohemia, the Elector Palatine, whose exiled court had been installed at The Hague under the protection of the Prince of Orange. The young princess, who was very intelligent and very witty, was particularly interested in mathematics and in metaphysics. Descartes found her a charming pupil and companion. An affectionate intellectual friendship soon sprang up between them and the level of the problems they discussed became so high that Descartes set his pupil the 'three-circles problem', i.e. the problem of constructing a circle tangent to three given circles. However, he soon felt a little worried at having done so. He wrote to M. Alphonse de Pollot on 21 October 1643:

I rather regret that I recently set the three circles problem to the Princess of Bohemia, since the problem is so difficult that I cannot believe even an angel who had no other instruction in algebra than that of St. [ampionen]* could solve it without a miracle'.

Descartes therefore thought it prudent not to wait to see how the Princess managed her task, and, to avoid offending her, prepared a detailed analytical solution which he sent to M. de Pollot with instructions that it should be given her, under certain conditions.

I thought myself obliged to send her the solution of the problem she believed she had solved, and to explain why I do not believe it can in fact be solved if one supposes there is only one root. Nevertheless, I hesitate to send her the solution, since she would perhaps prefer to continue to try to solve the problem for herself rather than read what I have written. If that is the case, please do not give her my letter too soon. I have not put any date on it. Perhaps she has, in fact, found the solution, but has not finished the calculations, which are long and tedious, and, in this case, I shall be pleased for her to see my letter, since it advises her not to do all the further work, which is quite unnecessary.

But Elizabeth had worked hard, and solved the problem, though she apologized for the fact that, not being sufficiently adept at Cartesian algebra, she had had to resort to ancient methods. Descartes, flattered and surprised, replied:

The solution which your Highness has been pleased to send me is completely satisfactory in every way, and seeing it I was not only amazed but also delighted and somewhat flattered to find that the method your Highness has used is exactly like the one I proposed in my *Geometry*.

At the beginning of the year 1649 Descartes was invited to Stockholm by Queen Christina. The visit proved disappointing. The life of the Court and the duties it imposed on him were burdensome: he was, for instance, required to be in the Library every morning at five o'clock, and was even obliged to compose a play and a ballet, in which the

---

* A joke at the expense of a rather boastful mathematician who had lost a stake of 600 florins in a bet with one of Descartes's pupils.

Queen danced. He suffered cruelly from the cold and from a sense of isolation. On 15 January 1650 he wrote: 'It seems to me that here men's thoughts freeze during the winter just as water does...; I promise you that my desire to return to my solitude increases every day more and more. I am not in my element here, and all I want is peace and quiet, blessings the most powerful kings on earth cannot give to those who do not know how to find them for themselves.'

Soon after writing these disillusioned words Descartes died, on 11 February 1650, after several days of violent suffering.

The great Dutch physicist, Christian Huygens, who was then only twenty-one years old, but had nevertheless impressed Descartes with his scientific merits, wrote a poem in French (though rather uncertain prosody) which shows his high regard for the great philosopher:

Epitaphe de Des Cartes par Chr. Huygens

Sous le climat glacé de ces terres chagrines,
Où l'hiver est suivi de l'arrière-saison,
Te voici sur le lieu que couvrent les ruines
D'un fameux bastiment qu'habita la Raison.

Par la rigueur du sort et de la Parque infame,
Cy gist Descartes au regret de l'Univers.
Ce qui servoit jadis d'interprete à son ame
Sert de matiere aux pleurs et de pâture aux vers.

Cette ame qui tousjours en sagesse feconde,
Faisoit voir aux esprits ce qui se cache aux yeux,
Après avoir produit le modele du monde,
S'informe desormais du mystere des cieux.

Nature, prends le deuil, viens plaindre la première,
Le Grand Descartes, et monstre ton desespoir,
Quand il perdit le jour, tu perdis la lumiére:
Ce n'est qu'à ce flambeau que nous t'avons pu voir!

Christ. Huygens, 1650

See translation on following page.

Descartes's *Geometry* which, with the treatises on *Dioptrics* (*Dioptrique*) and on *Atmospheric Phenomena* (*Météores*), was published together with the *Discourse on Method*, in 1637, is worthy of a place as a Classic alongside the last work.

It opens with a passage which marks an important stage in the development of science and is the first example of a modern mathematical style.

GEOMETRY

BOOK I

PROBLEMS WHICH CAN BE SOLVED USING ONLY CIRCLES AND STRAIGHT LINES

All geometrical problems can easily be reduced to terms such that at some later stage we only require to know the lengths of certain straight lines in order to construct solutions.

Translation from previous page.

Epitaph on Des Cartes by Christ. Huygens

Under the frozen climate of these unhappy lands
Where late autumn follows winter
You see the place which hides the ruins
Of a famous building once inhabited by Reason.

Harsh chance and the cruel Fates have brought
Descartes here, and the Universe mourns him.
What was once interpreter to his soul now provides
Matter for tears and food for worms.

That soul, always so rich in wisdom,
Showed spirits what was hidden from eyes,
And now, having explained the working of the world,
    turns from henceforth
To scan the mystery of the heavens.

Nature, put on mourning, come, be the first to weep
Great Descartes, and show your despair.
When his day ended you lost your light, for
Only this torch enabled us to see you.

Christ. Huygens, 1650
Trans. J.V.F.

How arithmetical calculations are related to geometrical operations

And just as arithmetic involves only four or five operations, namely addition, subtraction, multiplication, division and the extraction of roots (which can be taken as a kind of division) so, in geometry, in looking for the lines we require for our solutions we need only to add lines or subtract them; or we are given some line, which I shall call unity so as to facilitate the comparison with numbers (this unity can usually be chosen quite arbitrarily), and then, given two other lines, we are required to find a fourth line such that its ratio to one of these two is the same as that of the other to unity, which is the equivalent of multiplication; or we are required to find a fourth line whose relation to one of the two given lines is as unity is to the other, which is the same as division; or, finally, we have to find one, two or several means proportional between the unit line and some other line, which is the equivalent of taking a square or cube root, etc. I shall not hesitate to use these arithmetical terms in geometry in order to make my meaning clearer.

Multiplication

For instance, let *AB* be the unit and let it be required to *multiply BD* by *BC* [Fig. 10.4]. I merely join the points *A* and *C*, then draw *DE* parallel to *CA*, and *BE* is the required product.

Division

Or, if I am required to *divide BE* by *BD*, we join the points *E* and *D*, draw *AC* parallel to *DE*, and *BC* is the required quotient.

Extracting the square root

Or, if I am required to find the *square root* of *GH*, I produce *GH* by the line *FG*, the unit, and divide *FH* into two equal parts, find the mid-point of *FH*, which I shall call *K* [Fig. 10.5]. With centre *K* I draw the circle *FIH*, then construct at *G* a perpendicular to *FH* which cuts the circle at *I*, and *GI* is the required root. I shall not mention the cube or other roots here since it will be more convenient to deal with them later.

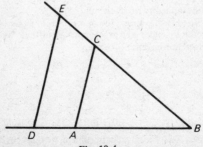

Fig. 10.4

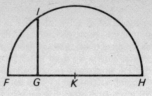

Fig. 10.5

How to use figures in geometry

But often there is no need to draw lines on the paper in this way. It is sufficient for us to designate each of them by a single letter. Thus, to add the line BD to GH, I call the lines a and b and write $a + b$, and $a - b$ for subtracting b from a, and ab for multiplying one by the other, and a/b for dividing one by the other, and aa or a^2 for multiplying a by itself and a^3 for multiplying once more by a, and so on indefinitely, and $\sqrt{aa + bb}$ for taking the square root of $aa + bb$, and $\sqrt{C \cdot a^3 - b^3 + abb}$ for taking the cube root of $a^3 - b^3 + abb$, and so on for the other operations.

It should be noted that when I use aa, or b^3, or similar notation, I do not ordinarily mean anything other than a straight line, although, in order to use the ordinary algebraical names, I refer to these quantities as squares, cubes, etc.

It should also be noted that all the parts of the same line should ordinarily be expressed as having the same number of dimensions as one another when the unit is not determined, just as, in the example given above, a^3 has the same dimensions as abb or b^3 in the expression I wrote for the line, namely

$$\sqrt{C \cdot a^3 - b^3 + abb};$$

but the same is not true when the unit is determined, because it can then be considered to be present wherever the dimensions are too large or too small, so that if, for example, we have to take the cube root of $aabb - b$, we must suppose that the quantity $aabb$ is divided once by the unit, and the quantity b is multiplied twice by the unit.

Moreover, so as not to forget the names that have been given to these lines, one should always make a separate list of them as the names are given or changed, for instance by writing:

$$AB \infty 1, \quad \text{that is to say } AB \text{ equals } 1$$

$$GH \infty a$$

$$BD \infty b, \quad \text{etc.}$$

How to arrive at the equations which are used to solve problems

So, if we wish to solve a particular problem, we must begin by supposing we have already carried out the required construction, and give names to all the lines which seem to be necessary for the construction, to the unknown ones as well as to the others. Then, without making any distinction between known and unknown lines, we must work over the figure, in the order in which we can see most clearly how the sizes of the lines depend on one another, and we must continue this process until we can express some quantity in two different ways, which will give us an equation, for the expressions obtained in these two different ways are equal to one another; and we need to find as many such equations as there were unknown lines in our construction. Otherwise, if we cannot find enough equations, and if we have nevertheless used all the information we were given in the problem, we have shown that the problem is not determinate. We can then take arbitrary values for all the unknown lines which do not give us equations. If this still leaves us with some lines, we must take each of the remaining equations in turn, either alone or with the others, to find values for each of these unknown lines, and then continue to manipulate them until we are left with an equation which either gives us the value of one of the lines directly or tells us that its square or its cube or the square of its cube or the fifth power of the line or the square of its cube, etc., is equal to some number obtained by the addition or subtraction of two or several other quantities of which one is known and the others are composed of some means proportional between unity and the square or cube or square of the square, etc., multiplied by other known quantities, which I write as follows:

$$z \infty b,$$

or

$$zz \infty -az + bb,$$

or

$$z^3 \infty +azz + bbz - c^3,$$

or

$$z^4 \infty az^3 - c^3z + d^4, \text{etc.},$$

that is, z, which I take as the unknown quantity, is equal to b or the square of z is equal to the square of b minus a multiplied by z, or the cube of z is equal to a multiplied by the square of z plus the square of b multiplied by z less the cube of c, and so on.

All the unknown quantities can always be expressed in terms of a single unknown quantity when the problem can be solved by a construction involving only circles and straight lines or also using conic sections or even using some other line constructed by only one or two further steps. But I shall not explain this in more detail at this point, since that would deprive you both of the pleasure of discovering it for yourself and also of the profit of improving your mind by making the attempt, which, in my opinion, is the chief profit to be gained from this study.

10.4. Desargues

Desargues qu'on peut appeler, à plus d'un titre,
le Monge de son siècle.*

Poncelet.

Girard Desargues, who was born in Lyon in 1593 and died in 1661, held an honourable position in the French scientific circles of his time. He was an engineer, or, more exactly, an architect, and his concern with the different techniques of his craft, the various forms of drawing (particularly in perspective), the cutting of stones, and the construction of sundials, led him to devise a completely new branch of mathematics: projective geometry. He was to have few immediate followers: apart from Blaise Pascal, who was interested in Desargues's work at the beginning of his scientific career, but unfortunately only for a very short time, there were Philippe de la Hire (1640–1718), a prolific and conscientious scholar, and the engraver, Abraham Bosse (1602–76).

The seed Desargues had sown did not really ripen until the end of the eighteenth century (in the work of Monge's pupils, Poncelet and Michel Chasles and the first half of the nineteenth century).

Desargues, like many original thinkers, was compelled to invent new words, and the number of new terms he employs makes his work difficult to follow. When Vieta had encountered a similar problem he took his new terms from the Greek, Desargues turned instead to a living language, the technical vocabulary of painters and masons (Plate XII).

Neither Vieta nor Desargues met with lasting success, but, on the other hand, a scientific vocabulary Stevin had then just recently invented for Flemish has, in the main, survived.

---

* Desargues who has more than one claim to be called the Monge of his century.

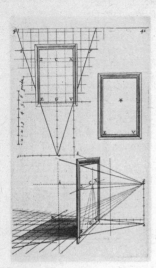

(a)   B.N. (b)   B.N.

PLATE XII. Illustrations from (a) *La manière universelle de M. Desargues pour pratiquer la perspective* (1648): and (b) *La pratique du trait à preuves de M. Desargues pour coup des pierres* (1643), both by A. Bosse.

Desargues stands at the beginning of projective geometry just as Descartes and Fermat, followers of Apollonius, stand at the beginning of analytical geometry.

The modern treatment of the geometry of conics is not directly derived from either of these two traditions, but rather from the work of Philippe de la Hire, which was partly based on that of Desargues.

Desargues's most important work, which is not easy to understand today, is the *Schematic sketch of what happens when a cone and a plane meet* (*Brouillon-Projet des événements de rencontre d'un cône avec un plan*, Paris 1639).

The following passage is translated from M. René Taton's modern edition of Desargues's work [2]:

When a straight line, constrained to pass through a fixed point, is moved round the edge or the circumference of a circle:

The fixed point of this straight line is either in the plane of the circle or not.

When the fixed point of the straight line is in the plane of the circle, it is either at a finite or an infinite distance from the circle.

And, whatever the position of this fixed point in the plane of the circle, when the straight line moves it always remains in the plane of the circle, and as it moves it defines a pencil [ordonnance] of straight lines which meet the circle and the vertex [but] of the pencil either at a finite or an infinite distance from the circle.

Roll [Rouleau]. When the fixed point of the line is outside the plane of the circle, it is either at a finite or an infinite distance from it, and, wherever the fixed point is outside the plane of the circle, when the straight line moves it always remains outside the plane of the circle, and in its movement it surrounds [environne], encloses [enferme] or describes [descrit] a solid figure which we shall call a roll, using this as a generic name which covers two subspecies.

Vertex [sommet] *of the roll*. The fixed point of the straight line is called the vertex of this roll.

Flat basis [assiette] *or Base* [base] *of the roll*. The circle round the edge of which the straight line moves is called the base or flat basis of the roll.

Envelope [envelope] *or Surface* [surface] *of the roll*. The space which the straight line sweeps out as it moves is called the envelope or the surface of the roll.

Column [colonne] *or Cylinder* [cylindre]. When the fixed point of the straight line is at an infinite distance from the plane of the circle around the edge of which it moves, the roll which it describes is of equal width at all points along its length which are at a finite distance from the circle, and we call it a column or cylinder, of which it is clear that columns or cylinders may be of various types.

Cornet [cornet] *or Cone* [cone]. When the fixed point of the straight line is at a finite distance from the plane of the circle, around the edge of which it moves, the roll which it describes by its revolution narrows towards the fixed point, at which its width is no more than that of a point, and on either side of this point it spreads out to infinity in two cornets whose vertices are opposed at this fixed point, which is why we call this roll a cornet, or otherwise a cone; it is clear that cornets or cones may be of various types.

Thus the column or cylinder and the cornet or cone are two sub-species of a type we have here called a roll. We shall for the most part consider in general case, and it will be clear that one part of this cornet or cone, a part which extends beyond the vertex on only one side, and which some people refer to as a complete cone, we should here consider to be only half a cornet or cone, and not a complete one.

And, finally, when we use this word cornet or cone we imagine the two parts together, the two parts opposed at their vertex, otherwise the cone is not a complete one.

A plane of section of the roll. When a plane other than that of the circle which forms the basis or base of the roll intersects this roll, this plane is called a plane of section of the roll.

A plane of section cuts a roll either at its vertex or not at its vertex, and, whenever it cuts it, it does so in such a way that when the straight line which describes the roll moves it is either never parallel to the plane of section or it is sometimes parallel to it.

10.5. Pascal

Incompréhensible. Tout ce qui est incompréhensible ne laisse pas d'être. Le nombre infini. Un espace infini égal au fini.*
Pascal (Pensées)

Pascal was born at Clermont on 19 June, 1623, into a family of lawyers. His father, Étienne Pascal, a notable scholar, who was on friendly terms with Roberval and Father Mersenne, took full responsibility for his son's education.

When his father went to live in Paris, the young Pascal was allowed to join in the learned discussions he had with his educated friends. At first he enthusiastically followed the new road that had been opened by Desargues, and at the beginning of 1640, when he was not yet seventeen years old, he published a brilliant though very short *Essay on Conics* (*Essay pour les Coniques*). It is in this pamphlet that we find the statement of Pascal's Theorem, in a form rather different from its modern one: 'The meets of the opposite sides of a hexagon inscribed in a conic are in a straight line.' The conjugate proposition, that 'in any hexagon circumscribed round a conic the lines joining opposite vertices are concurrent', which now seems an immediate consequence of Pascal's Theorem was, however, not discovered until about 1806 (i.e. about a century and a half later) by an Artillery Officer called Brianchon (1785–1864), though one of these

---

* Incomprehensible. What is incomprehensible nevertheless exists. Infinite number. Infinite space equal to a finite one.

theorems can be transformed into the other merely by using the elementary properties of poles and polars, which had been well-known to Philippe de la Hire in the seventeenth century.

The length of the interval between the discovery of the two theorems may perhaps be blamed on Pascal himself, since he seems to have turned away from projective geometry, though it might also be imputed to the fashion for employing analytical methods in geometry and to the fact that the differential and integral calculus was invented at the end of the century, it absorbed the interest of most of the geometers of the time.

For whatever reason, when Étienne Pascal left Paris for Rouen where, thanks to his return to Richelieu's favour, he was to hold high office, the energetic mind of his son had turned to matters other than Pure Geometry.

Firstly, he had become familiar with the experimental work of Torricelli, and, being a young man of good family, with plenty of leisure and adequate finances, he was able to develop his considerable talent for experimental Physics.

Secondly, he invented a calculating machine, which carried out additions in the manner of a modern cash register. This shows him to have been a competent engineer as well as a good businessman.

In the year 1654, Pascal wrote several papers concerning integers. The most important of these papers the *Treatise on the arithmetic triangle* (*Traité du triangle arithmétique*), deals with the coefficients of a binomial expansion.

This triangle, now often called *Pascal's triangle*, had been well known to algebraists for more than a century. We find it in the work of, among others, Stifel, Rudolff, Apianus, Tartaglia (Plate XIII), Jacques Peletier and Jean Trenchant. Trenchant calls it 'the trigon covered with symbolic numbers which generate one another according to a very important rule'.

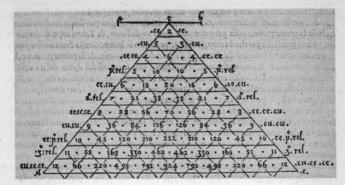

B.N.

PLATE XIII. The arithmetical triangle in Tartaglia's *General Treatise*.

Pascal must take the credit, not for having discovered the arithmetic triangle, but for having made a systematic and elegant study of it. In this paper, in particular, he makes frequent and happy use of the method of induction, which is peculiar to arithmetic. One of the results which he derives in this way is the formula we should write today as:

$$C_m^p = \frac{m(m-1)\ldots(m-p+1)}{1 \cdot 2 \cdot 3 \ldots p}.$$

We should note, however, that the formula itself was certainly known at least twenty-five years previously, in particular to Fermat.

The Chevalier de Méré had found several paradoxes in the theory of games of chance. These excited Pascal's interest, and led him to do some work which occasioned a fairly brief correspondence with Fermat.

In a letter dated 29 July 1654 Pascal says, among other things:

I have no time to send you the explanation of the point M. de Méré found difficult. He is very intelligent, but he is no geometer (which, as you know, is a serious deficiency) and cannot even understand that a mathematical line is infinitely divisible: he is quite sure a line is made up of a finite number of points, and I have never been able to convince him otherwise. If you can manage it, you would be doing him a great service.

Thus he was telling me he had found something wrong with numbers because:

If one is trying to throw a six* with a die, there is a better than even chance with four throws, by 671 to 625.

If one is trying to throw a pair of sixes with two dice, there is a worse than even chance with twenty-four throws.

Nevertheless the ratio of 24 to 36 (which is the number of the pairs of faces of the two dice) is the same as 4 to 6 (which is the number of faces of one die).

He found this very shocking and said loudly that the propositions were not consistent, and that arithmetic contradicted itself; but you are certainly learned enough to recognize the flaw in his reasoning.

Laplace, who was particularly interested in this sort of problem, makes the following remarks at the end of his *Philosophical Essay on Probabilities* (*Essai philosophique sur les Probabilités*).

HISTORICAL NOTE ON THE CALCULATIONS OF PROBABILITIES

In the simplest games it has long been the custom to work out players' chances of success or failure. Stakes and bets then were determined according to these chances. But before Pascal and Fermat nobody had described principles and methods for making such calculations, and only the simplest such problems had been solved. It was these two great geometers who laid the foundations of the theory of probabilities, a theory which takes its place among the noteworthy advances made in the seventeenth century, which was the time of the great achievement of the human spirit. The most important problem which they solved, by different methods, is that of making a fair division of the stake between players of equal skill who have agreed to stop playing before the end of a game, when in order to win a player had to be the first to get a certain number of points, a different number for each of the players. It is clear that the stake must be divided according to the probability that each player will win the game, these probabilities depending on the numbers of points they each require. Pascal's method is very ingenious, and basically consists of considering the partial differences involved, and using them to find the successive probabilities of each player's attaining his total, starting at the smallest numbers and working up.

---

* I.e. at least one six. H.G.F.

This method is limited to the case of two players. Fermat's method, which is based on combinations, can be extended to any number of players. Pascal at first thought that it was, like his own, restricted to two players, and this misunderstanding gave rise to an argument between them, at the end of which Pascal recognized that Fermat's method was general in its application.

Huygens collected together the various problems which had already been solved and added some new ones in a little treatise he wrote, the first treatise to be written on this subject: *De Ratiociniis in ludo aleae*. Later several geometers became interested: Hudde, the grand Pensionary, Witt in Holland, and Halley in England all applied this type of calculation to the probability of a man's attaining a certain age. Halley, in fact, published the first table of mortality figures. About the same time, Jacques Bernoulli devised various problems on probability for geometers, and later gave solutions to them. He wrote a fine work called *Ars conjectandi*, but it was not published until seven years after his death in 1705. In this work Bernoulli takes the theory of probabilities much further than Huygens did: he gives a general theory of combinations and series and applies it to several difficult problems concerning chance. The work is also noteworthy for the accuracy and subtlety of its insights, and for applying the binomial formula to this type of problem, as well as for giving a proof of the theorem that, if we increase, without limit, the number of observations and experiments, the ratio of the various outcomes tends towards the ratio of their respective probabilities, and by making the number of experiments large enough this ratio can be made as close as we please to the ratio of the probabilities. This theorem is very useful in helping us to deduce from observations the laws and causes associated with various phenomena. Bernoulli rightly attached great importance to his proof, which he claims to have thought out over a period of twenty years.

In the interval between Jacques Bernoulli's death and the publication of his work, Montmort and Moivre published two treatises on the calculations of probabilities. Montmort's is called *Essay on Games of Chance* (*Essai sur les Jeux de Hasard*), and it contains many applications of such calculations to various games. In the second edition the author added several letters in which Nicolas Bernoulli gives ingenious solutions to several difficult problems. Moivre's treatise, later than that of Montmort, first appeared in the *Philosophical Transactions* (*Transactions Philosophiques*) of the year 1711. The author later published it separately, and made successive improvements to it in the three editions which appeared. The work is mainly based on the binomial formula, and the problems it considers, as well as their solutions, are of great generality. But its most interesting feature is the theory of recurring series and the uses to which they may be put in problems of this sort. This is equivalent to integrating linear equations with finite differences and constant coefficients, an integration Moivre carries out very elegantly.

Moivre makes use of Jacques Bernoulli's theorem on the probability of results found from a large number of observations. Unlike Bernoulli, he is not content with showing that the ratio of the number of different outcomes continually approaches the ratio of their respective probabilities, but also gives an elegant and simple expression for the probability that the difference between these two ratios lies between given limits. To do this, he finds the ratio between the largest term of a binomial expansion of a very high degree and the sum of all the terms, and then finds the hyperbolic logarithm of the difference between the greatest term and the terms closest to it. Since the greatest term is the product of a considerable number of factors, it is not practicable to calculate it numerically. Moivre obtains a convergent approximation to it by using Stirling's theorem on the mean term of a binomial expansion of very high degree. This theorem is remarkable particularly in that it introduces the square root of the ratio of the circumference to the radius of a circle into an expression which seems quite unconnected with this transcendental number. Moivre was extremely interested by this result, which Stirling had deduced from an expression for the circumference of a circle in terms of an infinite product, an expression which Wallis had obtained by an interesting method of analysis which contains the germ of the fascinating and useful notion of definite integrals.

When Pascal was kept awake by toothache he took his mind off the pain by setting himself problems on a curve which had excited considerable interest during the previous twenty years: 'the roulette, otherwise called a trochoid or cycloid'.

The cycloid is so common that, after the straight line and the circle, there is no other curve which occurs so frequently; and everyone sees this curve so often that it is surprising that no mention of it is to be found among the works of the Ancients. For this curve is nothing more than the path of the nail in a wheel when the wheel rolls in the ordinary way, the path being taken from the point where the nail leaves the ground to the point where the continuous rolling of the wheel takes it back to the ground after one complete revolution, if we suppose the wheel to be a perfect circle, the nail a point on its circumference and the ground to be perfectly flat.

The rolling of the circle is considered in a Greek work called the *Mechanics*, which is ascribed to Aristotle but certainly dates from after his time, though probably from before the time of Euclid.

In particular, the *Mechanics* contains the following paradox (Fig. 10.6):

A circle with centre A and radius AB rolls along the straight line BB', and after one complete turn B comes to B'.

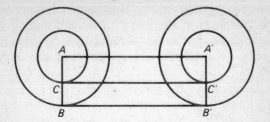

Fig. 10.6

This circle carries along another, concentric with it, having radius *AC*. *A*, *C* and *B* are in line. When *B* has come to *B'*, *C* is at *C'*. The straight segment *BB'* is equal to the circumference of the first circle. Is the circumference of the small circle, that is, *CC'* equal to *BB'*? If so, the two circles would have the same circumference.

This paradox had interested Galileo, who discussed it in his *Dialogues on the Two New Sciences* of 1638.

He was, moreover, also interested in the roulette itself, which he called a cycloid. Its elegance led him to recommend it as a suitable shape for the arches of bridges, but he had no success in treating it mathematically.

Roberval, following Mersenne's suggestion that he should study this curve, discovered that the area of the arch of a cycloid was three times the area of the generating circle. Once this result was known, in 1638, Fermat and Descartes each found a different way of deriving it. Roberval, Fermat and Descartes also found the tangent to the curve, and it was in the course of this work that Roberval developed his kinematic method of finding tangents.

In 1640, Father Nicéron, then travelling in Italy, suggested to Cavalieri that he should find the area of the cycloid. Cavalieri discussed the matter with Galileo and, on discovering that Galileo had made a lengthy but unsuccessful study of the curve, he decided not to engage upon what he took to be a hopeless task. He nevertheless mentioned the subject in his correspondence with Torricelli.

In the spring of 1643, Torricelli showed by a method of his own that the area under the arch of a cycloid was three times that of the generating circle, and at the same time, his pupil Viviani found the tangent using the kinematic method which Torricelli had deduced from Galileo's work and had himself applied to the Archimedean Spiral. At this point Ricci informed Torricelli that the French also knew the area of this curve, but Torricelli, convinced that the Italians had discovered it first, published his result in 1644 without mentioning the name of Roberval.

This resulted in a quarrel which lasted until 1647, the year of Torricelli's death.

Throughout this time, Pascal and his father were in Rouen, and did not follow the developments of the argument. The death of Father Nicéron (1646) and Father Mersenne (1648) did not make things any easier, and, if in his description of the cycloid Pascal seems over-inclined to blame Torricelli, he is not, in fact, as wilfully unjust as he has often been accused of being.

When Pascal himself came to study the cycloid the method of indivisibles had been developed. This helped him to find the areas of various parts of the arch of the curve, the positions of their centres of gravity and the volumes of revolution obtained when they were rotated round various axes.

In June 1658, encouraged by his friends, particularly the Duke of Roannez, Pascal, under the assumed name of Dettonville, issued a challenge to all mathematicians and offered a prize in kind to anyone who could succeed before October in solving the problems he proposed.

Such mathematical challenges were an old custom: we have mentioned them in connection with Leonardo of Pisa, in the thirteenth century, and in the sixteenth in connection with Cardan, Tartaglia and their fellow citizens. When Pascal issued his challenge an end had just been made to a polemical correspondence between Fermat and Frénicle on the one side and Wallis and Lord Brouncker on the other. This correspondence was published by Wallis in 1658 in the *Commercium Epistolicum*. The arguments concerned the theory of numbers, the equations later named after Pell but really invented by Fermat,* squaring the circle, squaring other areas, cubing volumes, finding centres of gravity when hyperbolic quantities were involved, etc. . . .

All established mathematicians were interested by the letters of the mysterious Dettonville, but only two responded to the challenge: Wallis and Lalouvère. John Wallis (1616–1703), a professor at Oxford, was a very good mathematician, and his solutions to some of the problems are better than those given by Pascal himself; but Pascal did not consider he had earned the prize.

Father Antoine de Lalouvère (1600–64), a Jesuit, was a very worthy professor of mathematics in the Jesuit College at Toulouse. Fermat's opinion of him was so high that he allowed some of his own papers to be published as appendices to Lalouvère's works.

Like many Jesuit professors, Lalouvère made conscientious and clumsy use of the methods of Archimedes, but nowadays he is, on the whole, no more difficult to read than Pascal on the same subject. He accepted the challenge only for the sake of honour and not for material reward. However, he came off very much the worse in his contest with his redoubtable opponent Dettonville-Pascal.

---

* We require to find two integers x and y such that $ax^2 + 1 = y^2$ where a is not a perfect square. This kind of equation is very important in number theory.

11. Classical Work

*C'était à peine l'aurore du beau jour que les
découvertes mathématiques de Newton, de
Leibnitz et de leur école préparaient à une
génération destinée à perfectionner ce que
ces grands hommes avaient commencé, et à
rétablir dans ses droits la raison si long-
temps étouffée sous le poids des préjugés.**

Lacroix

Modern mathematics appeared in its classical form at the
end of the seventeenth century, in the work of two men equal
in genius but very different in temperament: Isaac Newton
and Gottfried Wilhelm Leibniz.

Newton, a Cambridge don, the founder of modern
mechanics, was a scholar educated in the best scientific
tradition of the time.

Leibniz, thoroughly imbued with Scholasticism, a great
enthusiast for ideas, a historian, diplomat and philosopher,
was, in his mathematical work (like Pascal, but perhaps to
an even greater extent), an amateur of genius.

11.1. Newton

Isaac Newton, the son of a small landowner, was born
after his father's death, on Christmas Day, 1642, at Wools-
thorpe Manor in Lincolnshire. In 1661 he went up to

---

* The mathematical discoveries of Newton, Leibniz and their schools
merely marked the dawn of that brilliant age which was to see the achievement of what those great men had begun, when reason, so long stifled under
the weight of prejudices, was to be allowed once more to exercise her full
rights.

Trinity College, Cambridge, where he was taught by Isaac Barrow, then aged about thirty. Barrow was fluent in Greek and in Latin, and as familiar with ancient mathematics as with the work of his own time. Newton studied Kepler's *Dioptrice*, then the work of Descartes, Vieta, Oughtred (who had introduced Vieta's work into elementary mathematical teaching in England), and also that of van Schooten (who has popularized Descartes's work in the Netherlands). In 1663 he began to read the work of Wallis, the most important English mathematician of the time. His studies soon bore fruit. In a manuscript work of 1665 he stated the formula for a binomial expansion with any index and described the fundamental ideas underlying his method of fluents and fluxions, a method equivalent to the *calculus* of his rival Leibniz.

Newton's greatest work, the *Philosophiae naturalis Principia mathematica* (*Mathematical principles of natural philosophy*), was published in London in 1687.

This important treatise contains the statement of the law of universal gravitation and lays the foundations of classical mechanics, the principles of which were to dominate the physics of the eighteenth and nineteenth centuries until they were eventually replaced by the theory of relativity. The *Principia* also makes passing references to numerous geometrical and analytical ideas that were later to be of considerable importance.

This masterpiece was received rather cooly in France, as being too obviously in conflict with the accepted, Cartesian, ideas of physics. It was translated into French by the Marquise du Châtelet in 1759, with mathematical help from Clairaut, while Voltaire helped to spread Newton's ideas about physics and philosophy. Newton died on 20 March 1727.

In 1707 some of Newton's lectures were published, under the title *Arithmetica universalis*. The editor was Whiston, who had succeeded Newton in the professorship at Cambridge in which Newton himself had succeeded his teacher and friend, Isaac Barrow. Whiston's successor was to be the courageous blind mathematician, Saunderson. In one of the courses of lectures Newton gave between 1673 and 1683 he considered methods of solving equations [1]:

So far, I have developed properties of equations, explaining their trans-mutations, limits and reductions of every kind. Proofs I have not always attached since they have seemed easy enough to me and they could fre-quently not possibly be presented without excessive digression. It now remains merely to explain how to extract the roots of equations numerically once they have been reduced to their most convenient form. Here, the especial difficulty lies in obtaining their first two or three figures. That is most conveniently accomplished by some construction or other of the equation, be it a geometrical or a mechanical one. And for this reason it will not be an infliction to append certain constructions of the kind.

The Ancients, as we learn from Pappus, initially attacked the trisection of an angle and the finding of two mean proportionals by way of the straight line and circle, but to no effect. Subsequently, they began to take numerous other lines—such as the conchoid, cissoid and the conic sections—into consideration and by means of certain of these they solved those problems. At length, having pondered the matter more deeply and accepted conic sections into geometry, they distinguished problems into three types: plane ones, solvable by lines—namely, the straight line and circle—deriving their origin from the plane; solid ones, solved by lines deriving their source from consideration of a solid—a cone, to be exact—; and linear ones, for whose solution more complicated lines were required. Following this distinction it is alien to geometry to solve solid problems by any lines other than conics, especially if no other lines except the straight line, circle and conics are to be accepted into geometry. But mathematicians of more recent times have, in their further progress, welcomed into geometry all lines which can be expressed by means of equations, and have distinguished those lines into classes according to the dimensions of those lines, laying down the formal rule that it is not permissible to construct a problem by means of a line of a superior class when it can be constructed by one of a lower. In contemplating curves and deriving their properties I commend their distinction into classes in line with the dimensions of the equations by which they are defined. Yet it is not its equation but its des-cription which produces a geometrical curve. A circle is a geometrical line not because it is expressible by means of an equation but because its description (as such) is postulated. It is not the simplicity of its equation but the ease of its description which primarily indicates that a line is to be admitted into the construction of problems. To be sure, the equation to a parabola is simpler than that to a circle, and yet because of its simpler construction the circle is given prior admission. A circle and the conics are, if regard be paid to the dimensions of their equations, of the same order, and yet in the construction of problems a circle is not numbered with these latter curves but, because of its simpler description, is reduced to the lower order of the straight line; as a result it is not impermissible to construct by means of a circle what can be constructed by straight lines, but to construct by means of conics what can be constructed by a circle is to be reckoned a fault. Either, therefore, decree that the law (differentiating) by the dimensions of the equations is to be observed in the case of the circle and thus elide the distinction between plane and solid problems as being faulty, or allow that that law is not to be observed in lines of higher class in a way which precludes some, because of their simpler description, being given preference over others of the same order and, in the construction of problems, being ranked with lines of lower order. In constructions which are of equal geometrical rating the simpler ones are always to be preferred. This law overrides all exception. On the simplicity, indeed, of a construction the algebraic representation has no bearing. Here the descriptions of curves alone come into the reckoning. This consideration alone swayed the geometers who joined the circle with the straight line. According as these descriptions are easy or difficult the construction is rendered easy or difficult. It is hence alien to the nature of the subject to prescribe laws for constructions on any other basis. Either, then, we are, with the Ancients, to exclude from geometry all lines except the straight line and circle and maybe the conics, or we are to admit them all according to the simplicity of their description. Were the cycloid to be accepted into geometry, it would be allowable by its aid to cut up an angle in a given ratio. Could you then, if someone were to use this line to divide an angle in an integral ratio, see anything reprehensible in this and contend that this line is not defined by an equation, but that lines defined by equations need to be employed? We would in consequence, were the angle to be divided into (for instance) 10 001 parts, be compelled to bring into play a curve defined by an equation of more than a hundred dimensions: this, however, no mortal would be capable of describing, let alone comprehending and valuing above the cycloid—a curve which is exceedingly well known and very easily described through the motion of a wheel or a circle. How absurd this is, any one may see. Either, then, the cycloid is not to be admitted into geometry; or in the construction of problems it is to be preferred to all curves having a more difficult description. And the same reasoning goes for the rest of curves. On that head we commend the trisections of an angle by a conchoid (which Archimedes in his *Lemmas* and Pappus in his *Collection* placed above all the findings of others on this topic) inasmuch as one ought either to exclude from geometry all lines except the straight line and circle or to admit them according to the simplicity of their description, while the conchoid in the simplicity of its description yields to no curve except the circle. Equations are expressions belonging to arithmetical computation and in geometry properly have no place except in so far as certain truly geometrical quantities (lines, surfaces and solids, that is, and their ratios) are stated to be equal to others. Multiplications, divisions and computations of that sort have recently been introduced into geometry, but the step is ill-considered and contrary to the original intentions of this science: for anyone who examines the constructions of problems by the straight line and circle devised by the first geometers will readily perceive that geometry was contrived as a means of escaping the tediousness of calculation by the ready drawing of lines. Consequently these two sciences ought not to be confused. The Ancients so assiduously distinguished them one from the other that they never introduced arithmetical terms into geometry; while recent people, by confusing both, have lost the simplicity in which all elegance in geometry consists. Accordingly, the arithmetically simpler is indeed that which is determined by simpler equations, while the geometrically simpler is that which is gathered by a simpler drawing of lines—and in geometry what is simpler on geometrical grounds ought to be first and foremost. It will not therefore be interpreted as a fault in me if with the prince of mathematicians, Archimedes, and others of the Ancients I should employ a conchoid in the construction of solid problems. Nonetheless, if anyone does feel differently, I want him to know that my immediate concern is not for a construction which is geometrical, but for one of any sort whereby I may attain a numerical approximation to the roots of equations. With this motive I premise this lemmatical problem:

Between two given lines AB, AC to place a straight line BC of given length which, when produced, shall pass through the given point P.

11.2. Leibniz

Ce sçavant Géomètre a commencé où les autres avaient fini. Son Calcul l'a mené dans des pays jusqu'ici inconnus; et il y a fait des découvertes qui font l'étonnement des plus habiles Mathématiciens de l'Europe.[*]

L'Hôpital

Gottfried Wilhelm Leibniz was born on 21 June 1646, in Leipzig, where his father was a professor of Moral Philosophy. He enrolled as a student at Leipzig University in

---

[*] This accomplished geometer started his own work where others had ended theirs. His calculus led him into domains hitherto unknown and the discoveries he made amazed the most brilliant mathematicians of Europe.

1661, but spent a summer term at Jena, where he came under the influence of Edward Weigel (1625–99), the professor of Mathematics. Leibniz obtained his bachelor's degree in 1663, and in 1664 master's degrees in both philosophy and law. In 1666 he submitted his thesis 'Dissertatio de arte combinatoria' to the university of Altdorf.

During the years 1672 and 1675 Leibniz lived in Paris, as an attaché to the Embassy of the Elector Palatine, and from Paris he also visited London. He was lucky enough to meet some of the foremost mathematicians of the time, including Huygens, who had then just published his master-piece, an excellent treatise on the pendulum clock, the *Horologium oscillatorium*. This work, elegant and sober in style, makes use of methods taken from both pure and analytical geometry.

Huygens (1629–95), a man of unequalled mathematical scholarship, must be regarded as having been Leibniz's real teacher as far as mathematics was concerned.

Huygens recognized his young friend's brilliance, but he also realized he was lacking in mathematical education, so he encouraged Leibniz to read the works of Descartes, Gregory of St. Vincent, Pascal, Archimedes, Apollonius, and others.

The two great men with whom this chapter is concerned are the subjects of two parallel anecdotes: Newton is said to have realized that weight and the mutual attraction of heavenly bodies were manifestations of the same force when he saw an apple fall in his orchard; and Leibniz is said to have thought of the differential and integral calculus when he noticed the characteristic triangle in a work by Pascal. There is good reason to question the truth of both anecdotes, and in any case neither incident should be seen as more than the point of crystallization of a thought process during which a powerful intellect had unconsciously assembled the various elements of the problem.

According to Leibniz's own account, his methods were inspired mainly by Pascal's arithmetical work, particularly his tables of numerical values for the arithmetic triangle. So, paradoxically, the great philosopher who maintained that nature moves not by jerks but by continuous change treated curves and functions in a manner which made systematic use of the discontinuous, the discrete and the arithmetical. Above all, and in this he is considerably in advance of Pascal, Leibniz recognized the full importance of the progress that had been made by Vieta, though he came across Vieta's work only in the more highly developed form it had taken on in the writings of Descartes. Leibniz imagined a system in which the process of reasoning would be replaced by a set of rules that could in some respects be completely mechanical. This idea marks Leibniz as a forerunner of the advanced mathematicians of our own time. He was not able to carry out his scheme in full, but the vocabulary, symbolism, and algorithms he designed for the differential and integral calculus are still used today.

Leibniz was not much interested in geometrical rigour and he based his new work on a few simple principles which he did not attempt to justify. He is thus in some respects much inferior to Newton and Fermat, but both his notation and his approach to the subject were admirable and his methods therefore won the day.

The first book to bring Leibniz's work to the attention of the French public was the *Analysis of infinitely small quantities* (*Analyse des Infiniment Petits*) by Leibniz's friend and indirect pupil, Guillaume de l'Hôpital (1661–1704), who learned of the new methods from Jean Bernoulli. We shall not concern ourselves here with the originality of l'Hôpital's treatise, which is a model textbook, but shall merely quote some passages from it:

ANALYSIS OF INFINITELY SMALL QUANTITIES
PREFACE

The type of analysis we shall describe in this work presupposes an acquaintance with ordinary analysis, but is very different from it. Ordinary analysis deals only with finite quantities whereas we shall be concerned with infinite ones. We shall compare infinitely small differences with finite quantities; we shall consider the ratios of these differences and deduce those of the finite quantities, which, by comparison with these infinitely small quantities are like so many infinities. We could even say that our analysis takes us beyond infinity because we shall consider not only these infinitely small differences but also the ratios of the differences of these differences, and those of the third differences and the fourth differences and so on, without encountering any obstacle to our progress. So we shall not only deal with infinity but with an infinity of infinity or an infinity of infinities.

Only this kind of analysis is capable of giving a true insight into the properties of curves. For curves are merely polygons with an infinite number of sides, and curves differ from one another only because these infinitely small sides form different angles with one another. Only by using the methods of the analysis of infinitely small quantities can we determine the positions of these sides and thus find the properties of the curve they form: properties such as the directions of tangents and normals to the curve, its points of inflexion, its turning points, how it reflects or refracts rays, etc.

It has long been realized that polygons inscribed within a curve or circumscribed about it become identical with the curve as the number of their sides is increased to infinity. But there the matter rested until the invention of the type of analysis we are about to describe at last showed the scope and implications of such an idea.

The work done in this field by ancient scholars, particularly Archimedes, is certainly worthy of admiration. But they only considered a very few curves and those only cursorily. We find no more than a succession of special cases, in no particular order; and they provide no indication of any general and consistent method of procedure. We cannot reasonably blame the ancient scholars for this: it required a genius of the first order*

---

* Archimedis de lineis spiralibus tractatum cum bis térque legissem, totásque animi vires intendissem, ut subtilissimarum demonstrationum de spiralem Tangentibus artificium adsequerer; nusquam tamen, ingenuè fatebor, ab earum contemplatione ita certus recessi, quin scrupulus animo semper haereret, vim illius demonstrationis me non percepisse totam, etc. Bullialdus Praef. de lineis spiralibus. The ingenuous admission made by Ismaël Boulliau (1606–94) is, in its sincerity, nearer the truth than l'Hôpital supposes. Archimedes's treatise is not *without order*, but most seventeenth century mathematicians failed to understand him properly.

to find a way through so many difficulties and do the very first work in this hitherto completely unknown field. They did not go far, and they proceeded by roundabout ways, but, whatever Vieta says* they did not lose their way: and the more difficult and thorny the paths they trod the more we must admire these ancient mathematicians for having succeeded so well. To put it briefly: it does not appear that the Ancients could, at the time, have done better than they did. They did what our mathematicians would have done in their place, and if they were in ours it is likely they would do as we do now. All this is a consequence of the unchanging nature of the human mind and the fact that discoveries can only be made in an orderly succession.

It is therefore not surprising that the Ancients did not get any further; but it is a matter for great astonishment that great men, some assuredly as great as the Ancients, should have failed for so long to make any further progress, confining themselves to merely reading ancient authors and writing commentaries on their work, turning the knowledge they acquired to no use other than continuing to read, without daring to commit any such a crime as to think for themselves and look beyond what the Ancients had discovered. Thus it was that though there were many scholars, who wrote a great deal, so that the number of books multiplied, yet for all this activity there was no progress made: all the work of several centuries merely served to supply the world with respectful commentaries and repeated translations, often of quite uninteresting works.

Such was the state of mathematics, and, above all, of philosophy, until the time of M. Descartes, whose genius and self-confidence led him to abandon this study of ancient authorities and turn instead to reason, the authority to which these same Ancients had appealed. His boldness was considered to be merely a revolt but it led to any number of new and useful insights in physics and geometry. Then it was that men opened their eyes and began to think.

In mathematics, which is what concerns us here, M. Descartes began where the Ancients had left off; by solving a problem which Pappus said that no-one had been able to solve. It is well-known that he made great advances in analysis and in geometry, and that the techniques he evolved from combining the two make it possible to solve many problems which had previously seemed completely intractable. But since he was mainly concerned with the solution of equations he was interested in curves only

---

*Si verè Archimedes, fallaciter conclusit Euclides, etc. Supl. Geom. [In the passage quoted by l'Hôpital, Vieta makes some very pertinent remarks about Book V of the *Elements*, about squaring the circle, the angle of contact, etc. Vieta must have intended the sentence we have quoted to be taken slightly ironically, rather than quite literally. Historically, it is concerned with the dispute over the angle of contact which was carried on between Clavius on one side and Peletier and Vieta on the other.]

as a way to finding roots. He was therefore completely satisfied with using ordinary analysis, and did not consider the possibility of anything different. He did in fact use this new technique successfully in constructing tangents, and was so pleased with his method of solving this problem that he said it was 'the most useful and general problem he knew, or ever wanted to know, in all geometry'.

M. Descartes's *Geometry* made it fashionable to solve geometrical problems by means of equations, and opened up many possibilities of obtaining such solutions. Geometers applied themselves to the task and soon made new discoveries. Indeed, more and better new results are still being obtained.

M. Pascal was concerned with quite different matters: he studied curves as curves, and also as polygons. He found the lengths of some of them, the areas they enclosed, the volumes swept out by these areas, the centres of gravity of these areas and volumes, etc. etc. . . . By considering only elements, that is infinitely small portions of the curve, he discovered methods which were of general application, and which, moreover, we must find the more surprising in that he seems to have arrived at them merely by the power of his imaginative grasp and not by means of analysis.

Soon after M. Descartes had published his method for finding tangents M. Fermat discovered a different method, which M. Descartes himself finally admitted was in many ways better than his own. At the time, however, this method was not as simple as M. Barrow has since made it, by having paid closer attention to the properties of polygons, which naturally suggest that we consider the small triangles each made up of a part of the curve cut off between two infinitely close ordinates, the difference between these two ordinates and the difference between the corresponding abscissae. This triangle is similar to that formed by the tangent, the ordinate and the subtangent, so that this method of finding the tangent uses straightforward similarity instead of the calculations which were required by M. Descartes's method, and which this new method also previously required.

Barrow's work did not stop there: he also invented a kind of calculus based on this method, but this calculus, like that of Descartes, could only be used once all fractions and roots had been removed.

Barrow's calculus was replaced by that of Leibniz, an accomplished geometer who started his own work where Barrow and others had ended theirs. His calculus led him into domains hitherto unknown and the discoveries he made amazed the most brilliant mathematicians of Europe. The Bernoullis were the first to recognize the elegance of Leibniz's method, and they in turn developed his calculus to a degree which enabled them to solve problems which had previously seemed too difficult to attempt.

This calculus is of immense scope: it can be used for the curves which occur in mechanics [transcendental curves such as the catenary] as well as for purely geometrical curves, squares or other roots do not cause any difficulty (and may even be an advantage), any number of variables may be considered, and it is equally easy to compare infinitely small quantities of any type. An endless number of interesting results can be obtained concerning tangents (including tangent curves), concerning problems connected with maxima and minima, points of inflection and turning points, evolutes, caustics* derived by reflection or refraction, and so on, as we shall see in the work which follows.

I shall divide the work into ten sections. The first describes the principle of the calculus of differences.† The second describes how it is used to find the tangent to any curve, whatever the number of variables in the equation of the curve, though M. Craige‡ did not believe this method could be used for the transcendental curves which occur in mechanics. The third shows how the calculus is used in problems connected with maxima and minima. The fourth shows how it is used to find points of inflection and turning points. The fifth describes its use in finding M. Huygens's evolutes for all kinds of curves. The sixth and seventh sections show how it is used to find the caustic curves discovered by the distinguished scholar, M. Tschirnhaus, both the type formed by reflection and that formed by refraction. The method is again applicable to all kinds of curves. The eighth section describes how the calculus is used to find the curves which touch an infinite number of given straight lines or curves. The ninth consists of solutions to various problems arising out of the earlier work. The tenth section describes a new way of using the calculus of differences for geometrical curves: from which we can derive the method used by M. Descartes and M. Hudde, which is applicable only to this kind of curve....

I had intended to include an additional section which was to have described the marvellous use to which the calculus may be put in physics, what accuracy can thereby be attained, and to show how useful the calculus could be in mechanics. Illness however prevented me. The public will, nevertheless, not lose by this, since the work will eventually be published, with any additional material that has been accumulated in the meantime.

All this is only the first part of M. Leibniz's work on calculus, which consists of working down from integral quantities to consider the infinitely small differences between them and comparing these infinitely small differences with each other, whatever their type: this part is called the *Differential Calculus*. The other part of Leibniz's work is called the *Integral Calculus*, and consists of working up from these infinitely small quantities to the quantities or totals of which they are the differences; that is, it consists of finding their sums. I had intended to describe this also. But M. Leibniz wrote to me to say that he himself was engaged upon describing the integral calculus in a treatise he calls *De Scientia infiniti*, and I did not wish to deprive the public of such a work, which will deal with all the most interesting consequences of this inverse of the method of tangents,

---

* The *caustic* of a family of light rays is the 'envelope' of the family. The term was introduced by Tschirnhaus in 1682. H.G.F.

† Trans.: The modern term is 'differential calculus'.

‡ *Defigurarum curvilinearum quadraturis*, part 2. John Craig, an English mathematician (1660?–1731). The work in question dates from 1685.

showing how it can be used to find the lengths of curves, to find the area they enclose, to find the volumes and surface areas of their solids of revolution, to find centres of gravity etc. I have said as much as this only because M. Leibniz's wrote and asked me to do so, and I myself think it is necessary to prepare people's minds so that later they will be in a better position to understand all the results that are eventually obtained.

Finally, I am greatly indebted to the Bernoullis, particularly the younger Bernoulli, who is at present a professor at Groningen. I have made free use of their work as well as that of M. Leibniz. They may take the credit for as much of this work as they please, and I am quite content with what little they leave to me.

M. Leibniz himself acknowledges his debt to M. Newton, who, as it appears in his excellent work the *Philosophiae naturalis principia Mathematica* of 1687, had already invented a technique very like that of the differential calculus, which he uses throughout his book. But M. Leibniz's use of the *characteristic* makes his calculus much simpler and quicker, and sometimes also proves very helpful.

As the last pages of this book were being printed I came across M. Nieuwentiit's book. Its title, *Analysis infinitorum*, excited my interest, but when I read it through I found that it was very different from the present work: not only does the author not use M. Leibniz's characteristics but he also completely rejects second, third and further differences. Since he rejects what I have made the basis of most of my work I should feel obliged to reply to his objections, to show that they are unfounded, except that M. Leibniz has himself already made a more than adequate reply in the *Acta* of Leipzig*. Moreover, the two postulates or suppositions which I make at the beginning of this treatise, and which alone form the basis of what follows, seem to me to be so evidently true that no serious reader can reject them. I could, in fact, easily have proved them, in the manner of the Ancients, if I had not preferred to deal briefly with what was already well-known and enter into details only where the material itself was new.

SECTION I

DEFINITION I. We call *variables* those quantities which continually increase or decrease, and *constants* those quantities which remain the same while others change. Thus, for a parabola, the ordinates and the abscissae are variables whereas the latus rectum is a constant.

DEFINITION II. The infinitely small amount by which a variable continually increases or decreases is called its *difference*. ...

Note

In what follows we shall use the symbol or characteristic *d* to indicate the difference of the variable which itself is

---

* *Acta Eruditorum Lipsiensium.* J.V.F.

shown by a single letter. To avoid confusion the symbol d will not be used in any other sense in what follows ...

I.—Postulate or Assumption

2. Any two quantities may be replaced by one another if they differ from each other by no more than an infinitely small amount. . . .

II.—Postulate or Assumption

3. We may consider a curve as an assemblage of an infinite number of straight lines each infinitely short, or (equivalently) as a polygon with an infinite number of sides, each infinitely small, which, by the angles they make with one another, determine the shape of the curve. . . .

PROPOSITION II

Problem

5. To find the difference of the product of several quantities multiplied by one another.

First, the element of xy is $y\,dx + x\,dy$. For y increases to $y + dy$ when x becomes $x + dx$, so xy becomes $xy + y\,dx + x\,dy + dx\,dy$ (the product of $x + dx$ and $y + dy$) and its difference is therefore $y\,dx + x\,dy + dx\,dy$, that is (by Article 2) $y\,dx + x\,dy$, since $dx\,dy$ is infinitely small compared with the other two terms $y\,dx$ and $x\,dy$, for if, for example, we divide $y\,dx$ and $dx\,dy$ by dx we obtain y and dy, respectively, and the latter, being the difference of the former, is infinitely the smaller of the two. So the difference of the product of two quantities is equal to the product of the difference of the first multiplied by the second plus the product of the difference of the second multiplied by the first. . . .

SECTION II

The use of the calculus of differences to find the Tangents to all kinds of curves.

DEFINITION. If we produce the line Mm, one of the small sides of the polygon which forms a curve, this line will be called the *tangent* to the curve at the point M or m [Fig. 11.1].

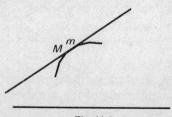

Fig. 11.1

PROPOSITION I

Problem

9. Let *AM* be a curve such that the relation between the length of the abscissa AP and the length of the ordinate *PM* is expressed by some algebraic equation [Fig. 11.2]. It is required to construct the tangent *MT* which touches the curve at a given point *M*.

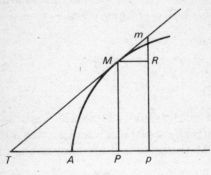

Fig. 11.2

We draw the ordinate *MP*, and suppose that the straight line *MT* which cuts the diameter at *T* is the required tangent. We then imagine another ordinate *mp*, infinitely close to the first one, and a small line *MR*, parallel to *AP*. Now, if we let $AP = x$ and $PM = y$ (so $Pp = MR = dx$ and $Rm = dy$) we have, from the similarity of triangles *mRM* and *MPT*, that $mR(dy) \cdot RM(dx) :: MP(y) \cdot PT = y\, dx/dy$. Now by considering the difference of the given equation we can obtain a value of dx in terms which all contain dy, and if we now multiply by y and divide by dy we obtain an expression for the subtangent *PT* entirely in terms of known quantities and free from differences, which enables us to draw the required tangent *MT*. . . .

Example I

11. First: if the relation between *AP* and *PM* is given by the equation $ax = yy$, then the curve *AM* will be a parabola having the given line *a* as semi-latus rectum. Taking the differences of both sides of this equation we obtain $a\, dx = 2y\, dy$ and thus

$$dx = \frac{2y\, dy}{a} \quad \text{and} \quad PT\left(\frac{y\, dx}{dy}\right) = \frac{2yy}{a} = 2x,$$

substituting ax for yy. Thus, if we take the point T such that PT is twice AP the line MT will be the tangent to the curve at the point M. Which is what we were required to construct.

Our story should really end at this point. But although we cannot trace the further development of mathematics we must still make some mention of those French scholars who were willing to devote some of their time and energy to the humble task of teaching others.

11.3. Clairault

*Votre guerre avec les Géomètres au sujet
de la comète me paraît la guerre des Dieux
dans l'Olympe.**

Voltaire to Clairaut

Alexis-Claude Clairaut was born in Paris on 13 May, 1713. His father, Jean-Baptiste Clairaut, was a distinguished teacher of Mathematics and a corresponding member of the Academy of Berlin [2].

In his éloge† on Clairaut, Grandjean de Fouchy gives an interesting description of his education:

'He was taught to recognize the letters of the alphabet from the diagrams in Euclid's *Elements*, the idea being that he would try to copy them and would want to understand how they were used. It was a sort of trap laid to catch his curiosity. It worked perfectly, and with the help of a few opportune rewards he learned to read, and to write quite well, by the time he was four years old.

'He also remembered the diagrams from Euclid's *Elements*, and mentioned them frequently, but before he could be allowed to study geometry he had to master arithmetic, a subject intrinsically less attractive than geometrical diagrams, particularly for a child. Again a similar sort of ruse

---

* Your war with the Geometers over the comet makes me think of the war among the Gods on Olympus.

† 'Eloges' were the printed texts of obituary speeches appearing in the *Mémoires* of the Paris Academy. H.G.F.

was employed: he was asked to write out all the integers, starting at one and going up to some very large number, putting each number into the position prepared for it. He was told that every time a number consisted entirely of nines the next number had to be written with as many zeros as there had been nines and a figure one had to be written to their left; also, some of the prepared positions were already filled with multiples of primes. This stimulated the child's interest, and in answer to his questions he learned more about the system of numerals than is known by many of the people who constantly make use of it. Multiplication and the other rules of arithmetic were introduced in a similar way, and Clairaut mastered this branch of mathematics almost without realizing he had studied it, or, at least, without finding his studies arduous.

'When he was nine ... he was given Guisnée's *Application of Algebra to Geometry* (*Application de l'algèbre à la Géométrie*). His father helped him to read it the first time, but he read it a second and a third time on his own, and it is said that after he had read the book for the third time he could already solve most of the problems in it more simply and more elegantly than the author of the book. Clairant's studies were already bringing out the inventiveness and lucidity of mind which were later to be his most notable qualities. He became so excited by the exercise of these new-found faculties that he had to be distracted from his work for fear he would damage his health. ...

'At the age of ten the young Clairaut began to read l'Hôpital's book on the conic sections. He managed to understand it, but found its principles rather harder to grasp than those of other books he had read. It seemed that he ought to read the book again, but he clearly disliked it, and all but refused to re-read it. Chance happened to intervene. M. de l'Isle, an Academician who was a close friend of Clairaut's father, came to see him, and found the young

Clairaut holding a copy of the Marquis de l'Hôpital's book. Not supposing that a child that age could possibly understand the work he said with a mocking smile that Clairaut was holding a book of which he had presumably read only the title and the cover. The boy was very annoyed by this insult. He managed to control himself, however, but the circumstances obliged him to read the book a second and even a third time. These re-readings, which he himself felt to be necessary, showed that while M. de l'Isle had not been quite justified in making his reproach he had not on the other hand been completely unjustified. Clairaut then read through l'Hôpital's *Analysis of Infinitely Small Quantities* and soon had a thorough grasp of the new methods and of the differential and integral calculus. . . .'

Clairaut was about twelve when 'his father moved to a new house, in which Clairaut and his younger brother were given a sort of study where they could work on their own. Unfortunately, this study was so placed that they could go in and out of it without being seen, and they did not fail to take advantage of the fact. They obtained a tinder-box, and when they were meant to be asleep they used to get up and spend most of the night working. Alexis, in the utmost secrecy, was engaged on writing a paper about the four cubic curves he had discovered, which could be used to find any number of means proportional between two given lines. He wanted the finished work to be a surprise but he was found out, being discovered at his task by his father, who strictly forbade him to indulge in such illicit studies. However, his father did not wish his work to have been done in vain so he introduced him to the Academy to read his paper. The paper suited so ill with the child's age that it was doubted whether he was indeed its author, and it was not until his answers to questions had proved that he was in fact capable of writing even more learned works that the company accorded Clairaut the praise that was due to him. Father Reyneau, who was present, was particularly moved, and could not contain tears of joy at seeing a child who was already worthy to be considered a great man.'*

It was at this time that Clairaut started work on skew curves. His famous paper on the subject, finished in 1729, but published only in 1731, was the first work to describe how to construct tangents to these curves. Clairaut made this important advance by using the techniques invented by Leibniz, which he had learned from studying the work of l'Hôpital.

'The curves we shall consider,' *wrote the eighteen-year-old author*, 'can only be drawn on the surfaces of curved solids: for example, the curves we should obtain if we were to use compasses to draw on the surface of a cylinder or some other curved surface. As far as I know no-one has dealt with this subject before. Descartes, who seems to have been the only person to have thought about this kind of curve, says that to study such curves we must draw perpendiculars from every point of the curve to two planes perpendicular to one another and relate every point of the curve to the points of the curves formed on these two planes.

I have considered the curves in this way throughout the present work, imagining them to be drawn within a solid quadrant rather as ordinary curves are seen as drawn within a plane quadrant. . . . I felt bound to call these curves 'curves with double curvature', because if we think of them in the manner described above we can see that they so to speak combine the curvatures of two curves. . . .'

Clairaut was elected to the Academy on 14 July 1731, and from then on he regularly attended its meetings and was one of its most active members. His work, which we shall not discuss here, was for the most part concerned with Celestial Mechanics and the Differential and Integral Calculus.

---

* Father Reyneau (1656–1728), of the Society of the Oratory, had taught for a long time at Angers. He was a friend of l'Hôpital and Malebranche and in his *Proof of Analysis* (*Analyse Démontrée*) and *The Science of Calculus* (*La Science du Calcul*) he described the mathematical ideas of Descartes, Newton and Leibniz. He regularly attended meetings of the Academy and 'paid great attention to all that was said there, which must have been the more difficult for him since he had for some time been rather hard of hearing'.

In 1741, however, he published his famous *Elements of Geometry* (*Eléments de Géométrie*), which was reprinted several times, and in 1746 there followed his *Elements of Algebra* (*Eléments de l'Algèbre*). The latter work in particular was to have a considerable influence on teaching in France. According to Lacroix 'Clairaut was the first to take up a definite philosophical position and give a lucid account of the principles of algebra'.

Clairaut died, exhausted by over-work, on 17 May 1765, at the age of only fifty-two.

11.4. D'Alembert

*C'est un des rare géomètres qui eut de l'esprit: Lisez ses Eloges.**

H. Bouasse

Jean le Rond d'Alembert was born in Paris on 16 November 1717 and died there on 29 October 1783.

His mother, the Marquise de Tencin, left him in a pine-wood box on the steps of the church of St. Jean le Rond, close to Notre Dame. He was taken into the Foundling Hospital, and for six weeks was put out to nurse in a village in Picardy. His father, the Chevalier Destouches, who had been away at the time of his birth, took him back from the Foundling Hospital, and placed him with a foster mother, Madame Rousseau, the wife of a glazier. He also bequeathed an income of 1200 livres (pounds) to the child.

D'Alembert attended a boarding school from the age of four until the age of twelve, when he went into the second form† of the Collège Mazarin, properly known as the Collège des quatre Nations.

This college then occupied what is now the site of the Institut. Mazarin founded it for the benefit of young noble-men from the four provinces which were conquered while

---

* He is one of the few witty geometers: read his éloges.

† Roughly equivalent to the sixth form of an English school today. J.V.F.

he was in power, namely Pignerol, Alsace, Flanders and Roussillon. It was one of the few colleges in Paris to have a professor of mathematics. In most other colleges, mathematics was, in fact, taught by the professor of philosophy, and generally only for one term in the two academic years of study. The Chair at the Collège Mazarin had been made famous by Varignon, and it was later given to a well-known astronomer, the Abbé de la Caille (1713–62).

When he left the college, d'Alembert studied law, but continued to be interested in mathematics. He himself wrote that

having no teachers, hardly any books and not even a friend he could ask for help when he was in difficulty, he used to go to public libraries, where he would read quickly in an attempt to grasp some general ideas and then would return home to try to work out proofs and solutions for himself. On the whole he succeeded, and even often discovered important propositions which he thought were new, only to be disappointed (and somewhat satisfied) when he came upon them later in books he had not read.

When he had finished studying law d'Alembert wanted to turn to medicine, and according to Condorcet 'he had his mathematics books taken to the house of a friend and did not intend to look at them until he had qualified in medicine, when he would be free to regard them as a recreation and not a distraction. His mind was still active, however, and from time to time he would ask his friend for a book he required to set his mind at rest when (and this is a feeling few men experience) he was tormented by the recollection of some result while unable to remember how to prove it. Gradually he took back all his books; and at last, convinced that resistance was useless, he yielded to his inclination, and committed himself to mathematics and to poverty. The years which followed this decision were the happiest of his life.'

In 1739 the Académie des Sciences showed interest in d'Alembert's *Memoir on the Integral Calculus*, which pointed out and corrected several important errors in Father Reyneau's *Proof of Analysis* (*l'Analyse Démontrée*). On 29 May 1741, at the age of 23, d'Alembert joined the Académie des Sciences as its astronomical correspondent, and there he met the famous Clairaut, a man slightly older than himself whom he was to dislike for the rest of his life.

D'Alembert played an important part in the intellectual life of the eighteenth century. His mathematical work is beyond the scope of this book, but one of his many interests was in improving the teaching of mathematics. His memoirs on this subject are to be found in his *Elements of Philosophy* (*Eléments de Philosophie*) and in his *Explanations of the Elements of Philosophy* (*Eclaircissements aux Eléments de Philosophie*).

His Explanation of the infinitesimal calculus reads as follows:

ON THE METAPHYSICAL PRINCIPLES OF THE INFINITESIMAL CALCULUS

To understand the branch of mathematics geometers refer to as infinitesimal calculus, we must first be very clear about what we mean by infinity.

If we think about it, we see that this idea is no more than an abstraction. We think of some finite extent, we take away the boundaries of this extent and we then have the idea of an infinite extent. It is in the same way, and only in this way, that we can think of an infinite number, an infinite time and so on.

From this definition, or rather from this analysis, it is clear how vague and imperfect an idea we have of what we mean by infinity: having, in fact, merely defined it as something indefinite (taking 'indefinite' to mean that the quantity is vague and has no boundaries, rather than merely that the boundaries of the quantity are not precisely fixed).

We also see that, for the purpose of analysis, an infinite quantity is really the limit of a finite one, that is, the finite quantity tends towards the infinite quantity, without ever reaching it, though it may be supposed to get closer and closer to it. This will, in fact, serve as an adequate geometrical and analytical definition of an infinite quantity, as the following example will show:

Let us consider the infinite series formed by the fractions $\frac{1}{2}, \frac{1}{4}, \frac{1}{8}, \frac{1}{16}$, etc., each term being half the previous one. Now, mathematicians say, and can prove, that if we suppose this series of numbers to continue to infinity its sum is equal to 1. This means, to put it in clearly defined terms, that the number 1 is the limit of the sum of the series of numbers, that is, the more terms of the series we consider, the closer their sum will be to 1, *and the sum can be made as close to 1 as we please*. This last condition is necessary for our definition of a *limit*. The number 2, for example, is not the limit of the sum of the series because, however many terms of the series we take, though their sum will indeed always be closer and closer to 2, it cannot approach 2 as closely as we please, since the difference between the sum and 2 will always be greater than unity.

In the same way, when we say that the sum of the series 2, 4, 8, 16 . . . , or of any other increasing series, is infinite, we mean that the more terms of the series we take the greater is their sum, and this sum can exceed any number however large. We must take this as our definition of infinity at least insofar as the term is used in mathematics. The idea it expresses is clear, simple and not open to different interpretations.

I shall not be concerned here to decide whether infinite quantities actually exist, whether space is really infinite, whether time is really infinite, or whether a finite amount of matter really contains an infinite number of particles. None of these questions has anything to do with the mathematical definition of infinity, which is, as I have said, that an infinite quantity is the limit of a finite one. For the purposes of mathematics it is not necessary to suppose that the limit really exists: it is quite sufficient that the finite never reaches it.

So geometry, while it does not deny that an actual infinity may exist, does not necessarily require us to suppose that infinity really exists. This is a satisfactory answer to a large number of the objections which have been raised to the idea of infinity in mathematics.

It has been asked, for example, whether some infinities are not greater than others, and whether the square of an infinite number is not infinitely greater than the number itself. The geometer's reply is a simple one: there is not necessarily any such thing as an infinite number, and the idea of an infinite number is merely an abstract idea which expresses an imagined limit which no finite number can ever reach.

In geometry, we can speak of second- and third-order infinities and can attach precise meanings to these terms without involving ourselves in obscure and contentious metaphysics. For example, if we say that *when a certain line becomes infinite, a certain other line which depends on it is second-order infinite*, this means that the ratio of the second line to the first, supposing both of them to be finite, increases as the length of the first line increases, and this ratio may be supposed to be greater than any finite number we may choose.

If we say that the second line is third-order infinite, we mean, in precise terms, that if we take the product of the second line multiplied by a finite line and consider the ratio of this product to the square of the first line, then this ratio increases as the first line increases and can be made greater than any given ratio.

In the same way, when we say that a curve is a polygon with an infinite number of sides, we mean that the curve is the limit of the polygons which can be inscribed within it or circumscribed about it; that is to say, the more sides these polygons have, the more nearly they are equivalent to the curve, and we can make them as close to it as we please by increasing the number of their sides.

In this way, we can attach clear, simple and precise meanings to expressions which make use of the term or the idea of *infinity*. Such expressions, which are very common in higher geometry, are like many others in this branch of mathematics—*expressions which seem inexact if taken in their apparent metaphysical sense and which must, in fact, be seen only as abbreviations which mathematicians have invented for the purpose of stating some fact which if described and stated in exact terms would have required many more words.*

What I have said about the infinitely great is also true of the infinitely small. The infinitesimal calculus does not suppose the existence of such quantities. We need to describe this idea in more detail.

Suppose, for example, that I require to find the tangent to a curve CAB at the point A [Fig. 11.3]. I first take two arbitrary points A and B on the curve and through them draw the straight line AB produced indefinitely in the directions of Z and X. This line, which clearly cuts the curve, I shall call a *secant*. I then draw some arbitrary fixed line CE in the plane of the curve, and from the two points A and B on the curve, I draw the ordinates AD and BE perpendicular to this fixed line CE, which I shall refer to as the axis of the curve. It is clear that the position of the secant is determined by DE, the distance separating these two ordinates, and by BO, the difference between their lengths, so that, if we knew the separation and the

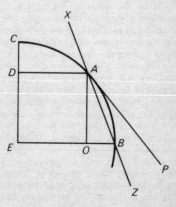

Fig. 11.3

difference, or even the ratio of the separation to the difference, we should know the position of the secant. Let us now imagine that of the two points A and B which we took to be on the curve one, for example the point B, slowly moves towards the other, the point A, and let us suppose that through this fixed point, A, we have drawn the line AP which is a tangent to the curve. It is easy to see that the secant AB, which is drawn through the two points A and B, slowly approaches the tangent to the curve as B approaches A, and will finally become identical with the tangent when the two points coincide. The tangent is thus the limit of the secants: they approach closer and closer to it, without ever reaching it, since they remain secants, though they may approach it as closely as we please. Now we have just seen that the position of the secant is determined by the ratio of BO, the difference between the lengths of the ordinates, to DE, the distance between them. So if we find the limit of this ratio, that is, the value it approaches more and more closely as the ordinates approach one another, this limit will give us the position of the tangent, since the tangent is the limit of the secants.

What, therefore, do we do in the differential calculus? We find the limit of the ratio of the finite difference between some two quantities to the finite difference between two other quantities, whose relation to the first two is given.

It is clear that the smaller each of these differences is the more closely does their ratio approach the required limit. It is also clear that so long as these differences are not actually zero, their ratio is not exactly equal to this limit, and when the differences are zero, the ratio can no longer be said to exist: for there is no ratio between two things which do not exist. But the limit of the ratio between these two differences, while they were still not yet zero, this limit is nonetheless real; and it is the value of this limit that, as we have seen, enables us to determine the position of the tangent.

An example may clarify what I have just said about the limit of the ratios. Let us consider two quantities such that the second is equal to twice the first plus the square of the first. It is clear that (1) the ratio of the second quantity to the first will always be greater than the number 2, so long as the first and second quantities are both non-zero; (2) that the ratio of the second quantity to the first will become closer to 2 as the first quantity becomes smaller, and that this ratio can be made as close to 2 as we please by taking the first quantity to be sufficiently small. From which it follows that the number 2 is the limit of the ratio of these two quantities; when the first of the two quantities becomes zero, the second obviously also becomes zero; and it is true to say that they then cannot be said to have any ratio to one another, but it is no less true, and no less obvious, that 2 is the limit of the ratio between the quantities while their values are still not zero.

Since the ratio of the differences approaches closer to its limit as the differences decrease in size, we suppose the limit of the ratio to be represented by the ratio of infinitely small differences. But once again this ratio of infinitely small differences is only a short way of expressing a more exact and more rigorous idea, that of the limit of the ratio of finite differences. For these infinitely small differences either do not really exist, or, at least, do not need to be supposed to exist in reality, in order that this limit shall be rigorously and exactly determined.

Some mathematicians have defined an infinitely small quantity as *a quantity 'which is vanishingly small, not before it vanishes, or after it has vanished, but at the very moment when it vanishes'*. I do not myself see what definite and precise idea such a definition is intended to express. A quantity exists or it does not. If it exists it has not yet vanished, and if it does not it has vanished completely. It is an illusion to suppose that there can be any third state between these two.

What we said above about different orders of infinity automatically applies to different orders of the infinitely small. When we say that a quantity is of the second order of smallness, that is, infinitely small compared with a quantity already infinitely small, we merely mean that the ratio of the first of these quantities to the second decreases as the second quantity decreases, and that the ratio may be made as small as we please by choosing a sufficiently small value of the second quantity.

In the same way, a quantity which is of the third order of smallness is such that, if it is multiplied by a finite quantity, the ratio of the resultant product to the square of another quantity decreases as this other quantity decreases, and can be made as small as we please.

It is clear from these principles how the differential calculus may be used to investigate the properties of curves. Differential calculus treats curves as the limit of polygons, and it is clear that the finite quantities whose ratios determine the properties of the polygons become zero for the curves, so that instead of determining the ratios of these quantities themselves the differential calculus determines the limit of their ratios, and thus provides us with a means of studying curves by considering them as the limits of polygons.

We can see from this that the differential calculus only, as it were, describes the properties of a curve at each of its points, since it merely enables us to calculate, at any point of the curve, the limit of the ratio of certain quantities which vanish for the curve but are finite for the polygon.

Differential calculus is the first branch of the infinitesimal calculus; the second is called *integral calculus*. We have just described the method of differential calculus. What does the *integral* calculus do? It gives us, in certain circumstances, a way of working back from the limit of the ratio of the differences between finite quantities to the ratio between these quantities themselves. By giving a value for this ratio it defines the curve as closely as it can be defined within a chosen finite range, since it enables us to inscribe a polygon inside the curve, or, equivalently, to find the properties of this polygon and the position of its sides.

Since all the problems to which the methods of differential and integral calculus can be applied are reducible to problems of finding curves and determining their properties, all that we have said about the philosophical basis of these types of calculus, and of their use in studying curves, is equally applicable to any other problem to which these same types of calculus can be applied.

The above therefore provides a sufficient introduction to the subject for those who merely wish to have a general, but correct, idea of its principles.

11.5. Lagrange, Laplace and Monge

We have seen that Clairaut and d'Alembert were both concerned with elementary mathematics.

After the French Revolution, the mathematicians of the next generation followed their example, and with much more powerful means at their disposal.

Condorcet's tragic death prevented him from helping his friends Monge, Lagrange, Laplace, Legendre and Lacroix in their task of planning how mathematics should be taught in France, a task they performed so well that France was to be the dominant influence in most of the mathematical work of the nineteenth century.

In secondary education, the founding of the Écoles Centrales, which were to become Lycées under the Empire, provided a tremendous impetus for the study of the exact sciences. The influence they exerted was mainly that of Lacroix and Legendre, and through them that of Clairaut and the school of Port-Royal.

Higher education was greatly stimulated by the founding of the École Centrale des Trauvaux Publics, which was soon to become the École Polytechnique. This was mainly due to Monge. It was, however, for the benefit of this institution that the aged Lagrange re-examined the whole of Higher Mathematics and thus played his part in bringing about that flowering of genius we find in the work of Cauchy or of Galois.

A little before the foundation of the École Centrale, the short-lived École Normale of the year III [1796] brought together a distinguished collection of teachers and students.

For instance, it was there that Monge first made public his work on the principles of descriptive geometry.

Fourier, who was a student at the school, has left us his recollections of the masters who taught there:

Laplace seems quite young; his voice is quiet but clear, and he speaks precisely, though not very fluently; his appearance is pleasant, and he dresses very simply; he is of medium height. His teaching of mathematics is in no way remarkable and he covers the material very rapidly. . . .

Monge has a loud voice and he is energetic, ingenious and very learned. It is well known that his talent is particularly for geometry, physics and chemistry. The subject he teaches is a fascinating one, and he describes it with the greatest possible clarity. He is even considered to be too clear, or, rather to deal with his material too slowly. He is to give individual practical lessons to his students. He speaks colloquially, and for the most part precisely. He is not only to be commended for his great knowledge but is also greatly respected in public and in private. His appearance is very ordinary. . . .

Lagrange, the foremost scholar of Europe, appears to be between fifty and sixty years old, though he is in fact younger; he has a strong Italian accent and pronounces an 's' as if it were a 'z'; he dresses very quietly, in black or brown; he speaks colloquially and with some difficulty, with the hesitant simplicity of a child. Everyone knows that he is an extraordinary man, but one needs to have seen him to recognise him as a great one. He only speaks in discussions, and some of what he says excites ridicule. The other day he said 'There are a lot of important things to be said on this subject, but I shall not say them.' The students are mainly incapable of appreciating his genius, but the teachers make up for it.

In the following passages from the reports of debates Lagrange begins by congratulating Laplace on his series of lectures on arithmetic, and the two teachers then reply to remarks and objections from students:

MEETINGS OF THE ÉCOLES NORMALES, NOTED DOWN IN SHORTHAND AND REVISED BY THE PROFESSORS

DEBATES

First Meeting (11 *Pluviôse*)*

Mathematics

Professors Lagrange and Laplace

---

* The fifth month of the Republican calendar (Jan.–Feb.) J.V.F.

Lagrange: Today we shall discuss Arithmetic; but before we begin I shall make a few remarks about some of the topics with which this subject is concerned. Since I have never written about Arithmetic I shall make my points in the order in which they occur to me.

You must have noticed, Citizens, that the easy, orderly and uniform way in which arithmetical operations can be carried out is a consequence of the fact that the number associated with each figure depends upon its position, the number being ten times greater each time the figure moves a place towards the left.

The idea of such an arrangement, simple as it is, nevertheless long eluded not only the general public but also the learned, including geometers. It first appeared in Europe only in the tenth century, when, it seems, a French monk called Gerbert learned of it from the Arabs, who then ruled Spain. Gerbret is reputed to have been the first to give currency to this idea, and to the rules of arithmetic that are its natural consequences. This is why our arithmetic is generally traced back to the Arabs.

The Ancients, for the most part, were only interested in the properties of individual numbers: and these properties do indeed raise many questions. There are theorems in number theory which are very difficult to prove, even more difficult than any known theorems in geometry and algebra: such as, for example, various theorems about prime numbers.

You have already been told that a number is said to be prime when it cannot be divided by any other number. Thus 3, 5, 11 and 13 are prime numbers. It is a strange thing that despite the efforts that have been made to find some rule for generating such numbers none has ever been found. We have, it is true, now discovered a million primes* but in order to do so we have had to test each time whether a particular number was divisible by another. It is true that the construction of tables made matters easier. There are now tables of prime numbers, which mainly serve to indicate which numbers are factors of others which are not prime.

It is often useful to know how a number can be expressed as a product of smaller numbers, that is, to know the factors of a given number. It provides us with a quick way of reducing a fraction to its lowest terms, by cancelling out the factors which occur in both the upper and the lower part of the fraction. This is a consequence of what is known in the theory of transformations as 'the chain rule'.

There are also tables which give the factors of various numbers, but such tables are not common, and they do not as yet go far enough to be very useful.

Up to the present time we have not found any *a priori* rule for recognizing that a number is prime, nor even a rule for generating prime numbers as

---

* When Lagrange said this, the efforts were on the point of achieving success. The French mathematician, A. Gérardin, who died in 1953, made a speciality of this study, and constructed a table which enabled numbers up to 200 million to be factorized. Large electronic computers can now go further still.

large as we please. There are, however, some rather elegant theorems concerning such numbers. For example, the following one, which does not in any way help us in our search for prime numbers but is very remarkable for its simplicity and generality:

If a number is prime (for example 5) the product of all the smaller numbers (2, 3 and 4), added to unity, will be divisible by the prime (5). In our example the product is 24. If we add 1 we obtain 25, which is divisible by 5. If we take 7 as the given prime, we find the product of 2, 3, 4, 5 and 6, which is 720, add 1, and obtain 721, which is divisible by 7.

If we take 11, we have the product of 2, 3, 4, 5, 6, 7, 8, 9 and 10 which is 3 628 800. Adding 1 we obtain 3 628 801, which is divisible by 11, the quotient being 329 891.

This theorem is one of the most elegant yet discovered. It is particularly interesting in that it is always true if the given number is prime and never true when the number is not prime, as can easily be verified.*

It is one of the advantages of our system of arithmetic that it allows us to handle fractions in the same way as integers.

The idea of decimals has already been described to you, but I think it is worth our while to discuss it further, since in the new system of weights and measures all the subdivisions are in terms of decimals. This will make all operations easier and will put an end to arithmetical operations with quantities which use several different bases, operations which are the bane of all young students of arithmetic.

There are very large books written about such operations, giving the particular rules for multiplying and dividing quantities such as livres [pounds] and onces [ounces], gros [drams] and grains [grains], and livres [pounds], sous [shillings] and deniers [pence], etc.

Since the subdivisions of these units do not all follow the same pattern, there were as many rules as there were different types of unit. For instance, taking the livre [pound] as unit we were told to take a tenth for two sous, for one sou half a tenth, for four sous a fifth, for five sous a quarter, etc.

All this complicated structure of rules is demolished if we use decimal arithmetic, since we can then handle fractions in the same way as integers.

In decimal arithmetic everything depends on the position of the decimal point, which the French indicate by a comma while the English and the Germans employ a full-stop. I prefer a comma to a full-stop; but, since a comma is often used to separate the figures of a number off by threes, it is, perhaps, better to avoid confusion by using a full-stop to indicate the position of the units. Once this position has been marked, arithmetical operations do not present any difficulty. In addition or subtraction we merely need to write the two numbers so that the decimal points are one

---

* Leibniz, in 1682, was the first mathematician to suspect the existence of the theorem Lagrange discusses here, but it was first published in 1770, by Waring, who credited John Wilson with having discovered it. It was first proved by Lagrange in 1773.

above the other, and then put a decimal point in this same position for the sum or difference. In multiplication we may treat the two numbers as integers, and then place the decimal point in the product in such a way that there are as many figures to the right of it as there are altogether to the right of the decimal points in the numbers we multiplied by one another, or we can put in the decimal point in the course of our operations, when we are multiplying units by units, since the product is to be expressed in units. In division, the quotient must contain a number of decimal places equal to the difference between the numbers of decimal places in the dividend and in the divisor. So the decimal point must be placed in the quotient in such a way that the number of figures to the right of it is the difference between the number of figures to the right of the decimal point in the dividend and the number of figures to the right of the decimal point in the divisor. Alternatively, we may put in the decimal point in the course of our operations, when we are dividing units by units, tens by tens, or hundreds by hundreds.

In general, since the decimal places in the quotient arise only from the difference in the number of decimal places in the dividend and the divisor we can, without changing the quotient, make equal changes in the number of decimal places in these two numbers, that is, we can move both their decimal points an equal number of places to the right or to the left. In this way we can always reduce the divisor to an integer, and the quotient will then contain as many decimal places as the dividend.

There is only one problem that arises in the use of decimal fractions: that for the most part their value is only approximate. In fact, the only fractions that can be expressed exactly in terms of decimals are those whose denominators are two, five, or some product of twos and fives, and no other factor. If we want to reduce any other fraction to a decimal by means of division, the process continues indefinitely, but we always reach a certain point after which the figures of the quotient begin to repeat themselves, in the same order, and then repeat themselves again and again to infinity. In fact, since the remainder is always less than the divisor it is clear that there cannot be more than a finite number of remainders, and that, once a remainder reappears, the division will give the same figures as before, and so on to infinity. We therefore obtain a quotient in which the same figures always recur. The French call this type of fraction periodic, and the English call it recurring. For example, if we wish to express $\frac{1}{3}$ as a decimal we obtain 0·3333 extended to infinity.

This seems to be inconvenient, and would indeed prove to be so if in everyday life we really demanded a rigorously mathematical standard of precision. But this is not so, for in any division we stop at a certain point: in money we do not go beyond the denier. For every purpose we observe a limit, and we only require to express that limit in terms of the units employed. Once this unit is fixed at the first, second or third decimal place we shall not need to go beyond that place in our calculations.

This solves the problem of the everyday use of decimal fractions.

We are limited by our senses, and there is thus a limit to everything.

It is only in science that we require a rigorous precision, but we do so mainly for intellectual satisfaction, to give, as it were, a definition of the goal which we must try to attain.

When we turn from theory to practice we are always obliged to content ourselves with more or less accurate approximations. And from this point of view we can say that a rigorous squaring of the circle and the finding of a general method of solving equations would have no practical advantages over the approximate methods we now use.

In any case, it is easy to find the exact value of a recurring decimal fraction: we need only consider the recurring part separately, and replace it by an ordinary fraction, whose numerator is made up of the figures which form the recurrent pattern and whose denominator is composed of an equal number of nines written one after the other.

For example, the fraction $0.333\ldots$ reduces to $\frac{3}{9}$, that is to $\frac{1}{3}$. The fraction $0.414141\ldots$ reduces to $\frac{41}{99}$. Now if we have the fraction $0.32414141\ldots$ where the recurrent pattern begins only at the third decimal place we should again substitute $\frac{41}{99}$ for the recurrent part, obtaining an exact value for the fraction of $0.32\frac{41}{99}$. We must note, however, that the fraction $\frac{41}{99}$ refers to hundredths, so it really represents $\frac{41}{9900}$.

The above procedure is clearly valid, since if we reduce the fractions $\frac{1}{9}$, $\frac{1}{99}$, $\frac{1}{999}$ to decimals (which we can do by division) we obtain the recurrent fractions $0.111\ldots$, $0.010101\ldots$ and $0.001001001\ldots$, a result we can also prove a priori.

Placiard: When you discussed the rules of arithmetic you spoke of addition, subtraction, multiplication and division. I noticed that the first three operations begin at the right whereas only the last one begins at the left. I should be grateful if you would explain the reasons why division is begun at the left rather than at the right. I suspect they are connected with the algebraic principle for ordering the powers of a given symbol.

Lagrange: That is a very good question. I admit I have thought about it myself more than once, and it seemed to me that, for the sake of uniformity at least, one ought also to begin subtraction at the left, since division is known to be a subtraction while multiplication is a repeated addition. Although one can, it is true, begin subtraction at the left it is less convenient that way.

As for division, it is clear it cannot be done any other way, since it must be begun in the opposite way to multiplication.

In multiplication we begin by multiplying the units, and then go on to the tens and the hundreds. In division we must invert the operation and start with the largest number.

That is why we begin the operation at the left. It is possible there are other reasons; I have thought about the question and have not found a satisfactory answer.

Laplace: I should like to add to my colleague's observations a remark of my own to the effect that the operations of arithmetic must be arranged so that, when they are performed, later operations do not affect the figures that have already been written down, a condition which is fulfilled by the methods currently employed. But it would not be fulfilled if the operations were carried out in the reverse order. For example, if we were to begin subtraction at the left we should take the figure on the left of the number which was to be subtracted and subtract it from the corresponding number above it and write the difference below. We should then move to the right and repeat the operation, but if the lower figure in that column were greater than the upper figure we should have to carry one from the first figure to the left of the number from which we were making the subtraction, and should thus need to decrease by one the figure we had already written for the difference.

The same inconvenience would be found in performing the other arithmetical operations if they were carried out in an order opposite to that usually adopted.

Lagrange: I should like to add one point. You saw that decimal fractions are obtained by division and that in almost every case the quotient continues indefinitely. The continuation is always towards figures of lower order, so it is necessary to begin the division at the side where there is a limit and continue it towards the side where it can continue indefinitely.

Placiard: I should also like to make an observation on the system of numbering.

Citizen Laplace said, in explaining the disadvantages of *binary* arithmetic, that one of the most important ones was that we should require a large number of characters to express a very simple number; that, for example, we should require eleven characters for 1024. I wanted to find out how to write 1024, and I found I had to perform eleven divisions. I began by dividing by 2, and then divided the *quotient* by 2, until I could no longer continue the process. When the *dividend* was smaller than the *divisor* I took 0 as *quotient* and the *dividend* as remainder: only the last figure was 1, the others were 0. I noticed that these remainders had to be written in the reverse order: the last was 1 and I had to reverse the remainders, writing first 1 and then to its right ten zeros. This unit gave me 1024 in *binary* arithmetic. I tried to extend this idea to *duodecimal* arithmetic, and thinking that the rule must be general I applied it to an arithmetic with base *A*. I had to reverse all the remainders and I noticed that these remainders, ordered in this way, did always give the correct expression for the number.

Laplace: In your journal you describe the rule for this procedure only for a *duodecimal* system. You have noticed, quite correctly, that it applies to all systems of numbering. If you remember what the journal says you will realize that the rule by which one writes the remainders down in the order they are obtained, each to the left of its predecessor, is equivalent to what you have just said.

A Student: Why is it that in the computation of logarithms the zero in the arithmetic progression is made to correspond to unity in the geometrical one?

Laplace: What is the purpose of logarithms? It is to reduce multiplication to additions and division to subtractions, and to simplify these operations as much as possible. This was done by making the zero in the arithmetic progression correspond to the unity in the geometrical one. We always have the same geometrical property: the ratio of unity to the multiplier is the same as the ratio of the multiplicand to the product; and to this geometrical relation there corresponds the arithmetic one: zero is to the logarithm of the multiplier as the multiplicand is to the logarithm of the product. Thus you can see that we obtain the logarithm of the product by adding the logarithm of the multiplier to the logarithm of the multiplicand. If we had not made the unity of the geometrical progression correspond to the zero of the arithmetic one we should have obtained the relation that the logarithm of unity is to the logarithm of the multiplicand as the logarithm of the multiplier is to the logarithm of the product, so to obtain the logarithm of the product from this arithmetic relation we should have been required to add the two central terms, the logarithm of the multiplier and the logarithm of the multiplicand, and then subtract the logarithm of unity, and this process would have to be carried out every time if the logarithm of unity were not zero.*

It is to avoid the necessity of carrying out this subtraction that in our tables of logarithms we make the logarithm of unity equal to zero, which is equivalent to making the unity of the geometrical progression correspond to the zero of the arithmetic one.

A Student: In weighing up the advantages and the disadvantages of decimal as opposed to duodecimal arithmetic, you discussed duodecimal arithmetic with some enthusiasm and described all its advantages.

However, after setting the advantages side by side with the disadvantages, you decided in favour of decimal arithmetic. Would it not have shown a certain amount of courage to have set yourselves up as legislators in this matter, for all nations would have followed you?

I am therefore asking you, Citizen, whether you really thought that duodecimal arithmetic was better.

Lagrange: Citizen, you are asking whether it would not in fact be advantageous to use duodecimal arithmetic instead of decimal arithmetic.

If we consider the question in the abstract much can be said on either side; but after all, since decimal arithmetic is universally employed, not only throughout Europe but also throughout the world, we can regard it as a kind of universal language, which it would be very inconvenient to change. If we were lucky enough to have for ordinary language, as we have

---

* Napier's original system of logarithms does in fact suffer from the disadvantage to which Laplace refers. Napier himself noted in an appendix to his *Constructio* how the system could be improved: 'Inter varios Logarithmorum progressus, is est praestantior, qui cyphram pro Logarithmo unitatis statuit....'

for numbers, one universal language, we should consider ourselves very fortunate, and it would not occur to anyone to wish to change it.

Considering the question theoretically I should say that: first of all I think duodecimal arithmetic has many advantages, because the number twelve has the advantage of being divisible by two, three, four, and six. With duodecimal numbers we should, therefore, be able to give exact expressions for a half, a third, a quarter and a sixth, and these fractions are so natural, and occur so frequently, that we slip into using them even when we do not intend to. I think this explains why in almost all the countries where decimal arithmetic is established we nevertheless find that for common and everyday purposes a duodecimal system is used. That is to say, people count in dozens. There are countries where people count in sixteens, and ancient astronomers adopted a sexagesimal system for their calculations, as being better than a decimal one because sixty has many more divisors than ten.

In the range from 1 to 24, twelve is the number which has most divisors, and in the range 1 to 120 the number with most divisors is sixty.

From the point of view of divisors the question is therefore answered; but I should like to make a further comment.

When we use decimal fractions the question of divisors and parts becomes irrelevant, and I shall show that for ordinary purposes it is much better to use decimal fractions than to use parts or ordinary fractions.

When we use numbers, the most important thing is that we should have a clear idea of their significance. When I say *one* I have the idea of single isolated thing, when I say *two* it is the same thing taken twice, for *three* it is the same thing taken three times, and so on.

It is not the same with fractions. They are much harder to visualize than integers. If I speak of a half I think of the same thing divided into two parts, if I speak of a third I must think of the same thing divided into three parts, visualizing one thing and mentally dividing into three parts. But when I come to compare fractions it is not an easy matter, and you will find that among those who have made a study of arithmetic there are few who could tell you at once how much greater a quarter is than a fifth. For example, you are asked for two and a third ells of cloth to make a suit. You think that a third is too much so you only get a quarter, but you have no definite idea of how much larger a third is than a quarter.

Fractions with different denominators, such as $\frac{1}{2}$, $\frac{1}{3}$ and $\frac{1}{4}$, though simple in themselves, are for that very reason less easy to use, because they are difficult to compare with one another. There are few people who could say at once how much greater a fifth is than a seventh. And you have seen, from what you have been told, that a certain amount of arithmetic is required if we want to express such fractions in terms of the same denominator. We only find it easy to compare fractional numbers which have the same denominator, since in that case we regard the denominator as a whole whose different parts are to be compared. This inconvenience is avoided by the use of decimal fractions, for the advantage of decimal fractions is that they always have the same denominator, or can easily be referred to it. So from *one* to *ten* we are only concerned with a single denominator and then again from 1 to 100, and so on. If, for example, you are asked for three metres and three decimetres of cloth and you find that it is not enough, you take four or five decimetres, etc., instead of three, and you always know how much more cloth you are taking, which you would not know if you employed ordinary fractions. If you consider centimetres and, for example, take three decimetres and five centimetres, although the denominator has now changed you can easily reduce your decimal fractions by calling your decimetres tens of centimetres. So you get thirty-five centimetres. In this way you can always use a constant denominator however small your decimal fractions may be.

It seems to me that this is a reason for always using decimal fractions for everyday purposes, because they enable one to form as clear an idea of fractional numbers as of integers.

We can see from the above that it does not matter really whether the number we use as a base (10 in the decimal system) has many divisors or not. There might even be advantages in using a prime number, such as 11 (which would be the base of a unidecimal system) because we should then be less inclined to employ fractions such as $\frac{1}{2}$, $\frac{1}{3}$ etc.

This concludes the authors' survey of what they have called 'classical work'. In the second volume they consider the history of certain mathematical methods and of some famous problems with which many mathematicians have been concerned over the centuries.   H.G.F.

Bibiographical references

Chapter 2

[1] Thureau-Dangin, *Textes Mathématiques Babyloniens* (Leyden, 1938), pp. 46–8, Pbs. 92–5.
[2] G. R. Morrow (tr.), *Proclus—a commentary on the first book of Euclid's Elements* (Princeton, 1970), pp. 124, 125.
[3] *Ibid.*, p. 195.
[4] Aristotle, *Analytica Priora*, I, 23.
[5] G. R. Morrow, *op. cit.*, XV.
[6] *Ibid.*, XXVI.
[7] Montucla, *Histoire des Mathématiques* (First edition, 1758).
[8] O. Neugebauer, *The Exact Sciences in Antiquity* (Copenhagen, 1951), p. 134.
[9] Heath, *Euclid's Elements III*, pp. 261, 262.

Chapter 3

[1] This information is taken from B. L. van der Waerden, *Science Awakening* (Groningen, 1954).
[2] Heath, *op. cit.*, p. 378.
[3] *Ibid.*, p. 382.
[4] *Ibid.*, p. 386.
[5] *Ibid.*, p. 392.
[6] *Ibid.*, p. 394.

Chapter 4

[1] Morrow, *op. cit.*
[2] This subdivision is taken from E. J. Dyksterhuis, *De Elementen van Euclides* (Groningen, 1921); and from Emanuel S. Cabrera, *Los Elementos de Euclides como exponente del 'milagro griego'* (Buenos Aires, 1949).
[3] Heath, *op. cit.* II, p. 114.
[4] *Ibid.*, p. 188.
[5] *Ibid.*, pp. 191–2.
[6] T. L. Heath, *A History of Greek Mathematics* (Oxford, 1921).
[7] *Ibid.*, II, pp. 128–30.

Chapter 5

[1] Quoted in François Woepcke *l'Algèbre d'Omar Al-Khayyâmi* (Paris, 1851).

Chapter 6

[1] P. Tannery, *Mémoires Scientifiques*, V, p. 317.
[2] Montucla, *op. cit.*

Chapter 8

[1] De Thou, *L'Histoire*, tr. A. Tessiet, vol. II, p. 400.
[2] J. Klein, *Greek Mathematical Thought* (Cambridge, Mass., 1968), tr. Eva Braun.

Chapter 10

[1] Most of the information and the quotations to be found in what follows are taken from the doctoral thesis of Gustave Cohen, who is now professor at the Sorbonne. The thesis is entitled *Ecrivains Français en Hollande dans la première moitié du XVIIe siècle* (Paris, 1920).
[2] *L'Oeuvre Mathématique de G. Desargues* (Paris, 1951).

Chapter 11

[1] D. T. Whiteside (ed. and tr.), *The Mathematical Papers of Sir Isaac Newton* (Cambridge, 1967).
[2] Biographical information about Clairaut's life is taken from Pierre Brunet, *La Vie et l'Oeuvre de Clairaut* (Paris, 1952).

Index of Names

This index is offered as a quick reference to the lives of the principal figures in the history of mathematics mentioned in this volume. Dates are given where known; where they are only roughly known, they are preceded by 'circa' (ca.); where they are doubtful, they are followed by a question mark; where only the century is known, this is stated, e.g. C.10 for tenth century. The page references are to biographical material: the references in italics are non-biographical.

313

General Index